Frontiers in Mathematics

Friedrich Kasch
Adolf Mader

Regularity and Substructures of Hom

Birkhäuser Verlag
Basel · Boston · Berlin

Authors:

Friedrich Kasch
Mathematisches Institut
Universität München
Theresienstrasse 39
D-80333 München
e-mail: Friedrich.Kasch@t-online.de

Adolf Mader
Department of Mathematics
University of Hawaii
2565 McCarthy Mall
Honolulu, HI 96822
USA
e-mail: adolf@math.hawaii.edu

2000 Mathematical Subject Classification: 08A05, 08A35, 13A10, 13C10, 13C11, 13E99, 16D10, 16D40, 16D50, 16S50, 20K15, 20K20, 20K25

Library of Congress Control Number: 2008939516

Bibliographic information published by Die Deutsche Bibliothek
Die Deutsche Bibliothek lists this publication in the Deutsche Nationalbibliografie; detailed bibliographic data is available in the Internet at <http://dnb.ddb.de>.

ISBN 978-3-7643-9989-4 Birkhäuser Verlag AG, Basel · Boston · Berlin

© 2009 Birkhäuser Verlag AG
Basel · Boston · Berlin
P.O. Box 133, CH-4010 Basel, Switzerland
Part of Springer Science+Business Media
Cover design: Birgit Blohmann, Zürich, Switzerland
Printed on acid-free paper produced from chlorine-free pulp. TCF ∞

ISBN 978-3-7643-9989-4   ISBN 978-3-7643-9990-0 (eBook)

9 8 7 6 5 4 3 2 1   www.birkhauser.ch

Contents

Preface

Regular rings were originally introduced by John von Neumann to clarify aspects of operator algebras ([33], [34], [9]). A continuous geometry is an indecomposable, continuous, complemented modular lattice that is not finite-dimensional ([8, page 155], [32, page V]). Von Neumann proved ([32, Theorem 14.1, page 208], [8, page 162]): *Every continuous geometry is isomorphic to the lattice of right ideals of some regular ring.* The book of K.R. Goodearl ([14]) gives an extensive account of various types of regular rings and there exist several papers studying modules over regular rings ([27], [31], [15]). In abelian group theory the interest lay in determining those groups whose endomorphism rings were regular or had related properties ([11, Section 112], [29], [30], [12], [13], [24]). An interesting feature was introduced by Brown and McCoy ([4]) who showed that every ring contains a unique largest ideal, all of whose elements are regular elements of the ring. In all these studies it was clear that regularity was intimately related to direct sum decompositions. Ware and Zelmanowitz ([35], [37]) defined regularity in modules and studied the structure of regular modules. Nicholson ([26]) generalized the notion and theory of regular modules.

In this purely algebraic monograph we study a generalization of regularity to the homomorphism group of two modules which was introduced by the first author ([19]). Little background is needed and the text is accessible to students with an exposure to standard modern algebra.

In the following, R is a ring with 1, and A, M are right unital R-modules.

Let $f \in \mathrm{Hom}_R(A, M)$. Then f is **regular** if there exists $g \in \mathrm{Hom}_R(M, A)$ such that $fgf = f$. If so, g is a **quasi-inverse** of f.

The isomorphisms $\mathrm{Hom}_R(R, M) \cong M$ and $\mathrm{Hom}_R(R, R) \cong R$ produce concepts of regularity in M_R and R which are exactly those used by earlier authors. The basic theme of this monograph is to generalize earlier results on rings and modules to homomorphism groups and to obtain new results on regularity in homomorphism groups that can then be specialized to rings and modules.

Chapter II contains basic properties of regular maps. Regular maps are associated with direct decompositions.

Corollary II.1.3 (Characterization of Regularity). $f \in \mathrm{Hom}_R(A, M)$ is regular if and only if $\mathrm{Ker}(f)$ is a summand A and $\mathrm{Im}(f)$ is a summand M.

This characterization already establishes that groups and rings of matrices over fields are regular since matrices are linear transformations and kernels and images of linear transformations are direct summands.

Special quasi-inverses exist producing special decompositions.

Proposition II.1.6. Let $f \in \operatorname{Hom}_R(A, M)$ be regular. Then there exists $g \in \operatorname{Hom}_R(M, A)$ such that $fgf = f$ and $gfg = g$. If so, then

$$A = \operatorname{Ker}(f) \oplus \operatorname{Im}(g), \quad \text{and} \quad M = \operatorname{Im}(f) \oplus \operatorname{Ker}(g).$$

Given modules A_R and M_R, let $S = \operatorname{End}(M_R)$ and $T = \operatorname{End}(A_R)$. Then $\operatorname{Hom}_R(A, M)$ is an S-T-bimodule and this is the setting in which we operate. Regular maps produce projective summands.

Theorem II.3.1. Let $0 \neq f \in \operatorname{Hom}_R(A, M)$ be regular. Then the following statements hold.

1) Sf is a nonzero S-projective direct summand of $_S\operatorname{Hom}_R(A, M)$ that is isomorphic to a cyclic left ideal of S that is a direct summand of S.

2) fT is a nonzero T-projective direct summand of $\operatorname{Hom}_R(A, M)_T$ that is isomorphic to a cyclic right ideal of T that is a direct summand of T.

There exists a largest regular S-T-submodule of $H = \operatorname{Hom}_R(A, M)$, denoted by $\operatorname{Reg}(A, M)$. Here "largest" means that any other regular S-T-submodule of H is contained in $\operatorname{Reg}(A, M)$.

Theorem II.4.3. $\operatorname{Reg}(A, M) = \{f \in \operatorname{Hom}_R(A, M) \mid SfT \text{ is regular}\}$ is the largest regular S-T-submodule of $\operatorname{Hom}_R(A, M)$.

Previous authors studied regular rings and modules and modules over regular rings. We are now confronted with the more delicate problem of computing $\operatorname{Reg}(A, M)$, and the corresponding specializations $\operatorname{Reg}(M_R)$, $\operatorname{Reg}(R)$, and $\operatorname{Reg}(A, A) = \operatorname{Reg}(\operatorname{End}(A_R))$.

There are interesting results on the structure of $\operatorname{Reg}(A, M)$.

Theorem II.4.6. Every finitely or countably generated S-submodule of $\operatorname{Reg}(A, M)$ is a direct sum of cyclic S-projective submodules that are isomorphic to left ideals of S and these ideals are direct summands of S.

Every finitely generated S-submodule L of $\operatorname{Reg}(A, M)$ is S-projective and a direct summand of $_S\operatorname{Hom}_R(A, M)$.

The analogous results hold for $\operatorname{Reg}(A, M)$ as a right T-module.

Proposition II.2.5. Every epimorphic image of $_S\operatorname{Reg}(A, M)$ is S-flat, and every epimorphic image of $\operatorname{Reg}(A, M)_T$ is T-flat.

Corollary II.4.8. Suppose that $\operatorname{Reg}(A, M)$ contains no infinite direct sums of S-submodules. Then $\operatorname{Reg}(A, M)$ is the direct sum of finitely many simple projective

S-modules, $\mathrm{Hom}_R(A, M) = \mathrm{Reg}(A, M) \oplus U$ for some S-submodule U and U contains no nonzero regular S-T-submodule. Every cyclic S-submodule of $\mathrm{Reg}(A, M)$ is isomorphic to a left ideal of S that is a direct summand of S.

Corollary II.4.9. Suppose that $\mathrm{Hom}_R(A, M)$ is regular. Then every finitely generated S-submodule L of $\mathrm{Hom}_R(A, M)$ is S-projective and a direct summand of $_S\,\mathrm{Hom}_R(A, M)$. Furthermore, L is the direct sum of finitely many cyclic S-projective submodules that are isomorphic to left ideals of S.

There is a connection between regular elements in $\mathrm{Hom}_R(A, M)$ and regular elements in $\mathrm{Hom}_R(M, A)$.

If $f \in \mathrm{Hom}_R(A, M)$ and $g \in \mathrm{Hom}_R(M, A)$ with $fgf = f$, then we call (f, g) a **regular pair**. Similarly, if $h \in \mathrm{Hom}_R(M, A)$ and $k \in \mathrm{Hom}_R(A, M)$ with $hkh = h$, then (h, k) is a regular pair. If (f, g) is a regular pair, then also (gfg, f) is a regular pair. We show that the so-called **transfer** $(f, g) \mapsto (gfg, f)$ produces all regular elements in $\mathrm{Hom}_R(M, A)$ from those in $\mathrm{Hom}_R(A, M)$.

We study "inherited regularity" when maps $\phi : A \to A'$ and $\mu : M \to M'$ are given and induce mappings on the homomorphism groups. The really important case of "inherited regularity" occurs when $A = A_1 \oplus \cdots \oplus A_m$ and $M = M_1 \oplus \cdots \oplus M_n$. We study the relationship between regular elements of $\mathrm{Hom}_R(A, M)$ and those of its additive subgroups $\mathrm{Hom}_R(A_i, M_j)$. It is convenient to identify $\mathrm{Hom}_R(A, M)$, $\mathrm{End}(A_R)$ and $\mathrm{End}(M_R)$ with groups and rings of matrices.

The results for $m = n = 2$ are as follows.

Theorem II.6.14. Let $A = A_1 \oplus A_2$ and $M = M_1 \oplus M_2$ as before. We use the identifications

$$\mathrm{Hom}_R(A, M) = \left\{ \begin{bmatrix} \xi_{11} & \xi_{21} \\ \xi_{12} & \xi_{22} \end{bmatrix}, \begin{array}{l} \xi_{11} \in \mathrm{Hom}_R(A_1, M_1), \quad \xi_{21} \in \mathrm{Hom}_R(A_2, M_1), \\ \xi_{12} \in \mathrm{Hom}_R(A_1, M_2), \quad \xi_{22} \in \mathrm{Hom}_R(A_2, M_2). \end{array} \right\},$$

$$S = \mathrm{End}_R(M) = \left\{ \begin{bmatrix} \mu_{11} & \mu_{21} \\ \mu_{12} & \mu_{22} \end{bmatrix} \mid \mu_{ij} \in \mathrm{Hom}_R(M_i, M_j) \right\}$$

and

$$T = \mathrm{End}_R(A) = \left\{ \begin{bmatrix} \alpha_{11} & \alpha_{21} \\ \alpha_{12} & \alpha_{22} \end{bmatrix} \mid \alpha_{ij} \in \mathrm{Hom}_R(A_i, A_j) \right\}.$$

Then

$$\begin{bmatrix} \xi_{11} & \xi_{21} \\ \xi_{12} & \xi_{22} \end{bmatrix} \in \mathrm{Reg}(A, M)$$

if and only if for all $\mu_{jt} \in \mathrm{Hom}_R(M_j, M_t)$ and for all $\alpha_{si} \in \mathrm{Hom}_R(A_s, A_i)$

$$\mu_{jt} \xi_{ij} \alpha_{si} \in \mathrm{Reg}(A_s, M_t).$$

Theorem II.6.14 can be extended to arbitrary finite sums by induction. There is an interesting corollary.

Corollary II.6.18. Let $A = A_1 \oplus \cdots \oplus A_m$ and $M = M_1 \oplus \cdots \oplus M_n$. Then $\mathrm{Hom}_R(A, M)$ is regular, i.e., $\mathrm{Reg}(A, M) = \mathrm{Hom}_R(A, M)$, if and only if

$$\forall i, j : \mathrm{Hom}(A_i, M_j) = \mathrm{Reg}(A_i, M_j).$$

In Chapter III we consider the case that either A or M is indecomposable and $\mathrm{Reg}(A, M) \neq 0$. If this is the case much can be said about the structure of both modules.

Theorem III.1.1.

1) Suppose that there is $0 \neq f \in \mathrm{Hom}_R(A, M)$ that is regular and μf is regular for all $\mu \in S = \mathrm{End}(M_R)$. If M is directly indecomposable, then S is a division ring.

2) Suppose that there is $0 \neq f \in \mathrm{Hom}_R(A, M)$ that is regular and $f\alpha$ is regular for all $\alpha \in T = \mathrm{End}(A_R)$. If A is directly indecomposable, then T is a division ring.

Corollary III.2.3. Suppose that $0 \neq f \in \mathrm{Reg}(A, M)$, and M is indecomposable. Then $\mathrm{End}(M_R)$ is a division ring and either $A = K \oplus M_1 \oplus \cdots \oplus M_n$ where $M_i \cong M$, and $\mathrm{Hom}_R(K, M) = 0$ or for every $i \in \mathbb{N}$ there is a decomposition $A = K_i \oplus M_1 \oplus \cdots \oplus M_i$ with $M_i \cong M$ and $K_i = K_{i+1} \oplus M_{i+1}$. In the latter case $\sum_{i=1}^{\infty} M_i = \bigoplus_{i=1}^{\infty} M_i$ and there is a homomorphism $\rho : A \to \prod_{i=1}^{\infty} M_i$ such that $\bigoplus_{i=1}^{\infty} M_i \subseteq \rho(A)$ and $\mathrm{Ker}(\rho) = \bigcap_{i=1}^{\infty} K_i$.

There is a similar structure theorem when A is indecomposable (Corollary III.2.7).

In Chapter IV we specialize the general theory to the case $\mathrm{Hom}_R(R, M) \cong M$. Then M is an S-R-bimodule where $S = \mathrm{End}(M_R)$. An element $m \in M$ is **regular** with **quasi-inverse** $\varphi \in \mathrm{Hom}_R(M, R)$ if $m = m\varphi(m)$. This means that $f \in \mathrm{Hom}_R(R, M)$ *is regular in* $\mathrm{Hom}_R(R, M)$ *if and only if* $f(1) \in M$ *is regular in* M.

Regular modules are special as the following example shows.

Theorem IV.1.2. Let $M \neq 0$ be a regular module over an integral domain R. Then R is a field. A vector space over a division ring is regular.

By means of the isomorphism $\mathrm{Hom}_R(R, M) \cong M$ we can specialize our general results to immediately obtain numerous results on regularity in modules (Theorem IV.1.3, Theorem IV.1.4, Theorem IV.1.8, Theorem IV.3.1, Theorem IV.3.3, Theorem IV.3.4, Theorem IV.3.7).

In Chapter V we consider $H = \mathrm{Hom}_R(A, M)$ as a bimodule but also as a one-sided module. Let $S = \mathrm{End}(M_R)$ and $T = \mathrm{End}(A_R)$. Then we have the bimodule $_S H_T$ and the one-sided modules $_S H$ and H_T and for each of these structures their concepts of regularity. Let $f \in H = \mathrm{Hom}_R(A, M)$.

- f is **S-regular**, i.e., f is regular as an element of $_SH$, if and only if there exists $\sigma \in \mathrm{Hom}_S(H, S)$ such that $(f\sigma)f = f$.

- f is **T-regular**, i.e., f is regular as an element of H_T, if and only if there exists $\tau \in \mathrm{Hom}_T(H, T)$ such that $f(\tau f) = f$.

It turns out that the regularity of $f \in \mathrm{Hom}_R(A, M)$ implies all other kinds of regularity.

Lemma V.2.1. Let $f \in \mathrm{Hom}_R(A, M)$ be regular and let $g \in \mathrm{Hom}_R(M, A)$ be such that $fgf = f$. Then the map

$$\sigma : {}_S\mathrm{Hom}_R(A, M) \ni h \mapsto (h)\sigma = h \circ g \in {}_S\mathrm{End}(M_R) = {}_SS$$

is a well-defined S-homomorphism and f is S-regular with quasi-inverse σ. The map

$$\tau : \mathrm{Hom}_R(M, A)_T \ni h \mapsto \tau(h) = g \circ h \in \mathrm{End}(A_R)_T = T_T$$

is a well-defined T-homomorphism and f is T-regular with quasi-inverse τ.

There are numerous immediate corollaries to our results on regularity for modules.

The existence of various largest regular submodules is an immediate consequence of Theorem II.4.3 and there are structure theorems that hold for the modules $_SH$ and H_T.

Theorem V.3.4. Let $H = \mathrm{Hom}_R(A, M)$, let $S' = \mathrm{End}(H_T)$ and suppose that N is a finitely generated S'-submodule of $\mathrm{Reg}(H_T)$. Then N is an S'-projective direct summand of H. Furthermore, N is the direct sum of finitely many cyclic projective submodules, each of which is isomorphic to a left ideal of S'. The same is true for submodules of $\mathrm{Reg}(H_T)$ considered as a right T-module.

In Chapter VI we introduce generalizations of regularity such as U-regularity and semiregularity. Semiregularity for modules was introduced by W.K. Nicholson ([26]). Zelmanowitz studied the class of regular modules, and Nicholson was interested in the wider class of semiregular modules. We list without proof some of his more striking results. Semiregularity is then defined for Hom and there are three different possibilities: semiregular, S-semiregular and T-semiregular. A sample result is as follows.

Theorem VI.4.2. If $f \in \mathrm{Hom}_R(A, M)$ is semiregular, then Sf lies over an S-projective direct summand of $_S\mathrm{Hom}_R(A, M)$.

In Chapter VII we consider additional substructures of $H = \mathrm{Hom}_R(A, M)$ and study relations between them.

The **singular submodule** of $_SH_T$ is by definition the S-T-submodule $\Delta(A, M) = \{f \in \mathrm{Hom}_R(A, M) \mid \mathrm{Ker}(f) \text{ is large in } A\}$. The **cosingular submodule** of H is the S-T-submodule $\nabla(A, M) = \{f \in \mathrm{Hom}_R(A, M) \mid \mathrm{Im}(f) \text{ is small in } M\}$, and there are three different concepts of radical in H: the radical of $_SH$ as an S-module, and similarly the radical of H_T as a T-module, and finally the most important radical for which there are two equivalent definitions: $\mathrm{Rad}(\mathrm{Hom}_R(A, M)) = \mathrm{Rad}(A, M) = \{f \in H \mid f\, \mathrm{Hom}_R(M, A) \subseteq \mathrm{Rad}(S)\} = \{f \in H \mid \mathrm{Hom}_R(M, A)f \subseteq \mathrm{Rad}(T)\}$.

Proposition VII.2.4. If A or M is large restricted or injective, then $\Delta(A, M) \subseteq \mathrm{Rad}(A, M)$.

Corollary VII.3.1.

1) If R_R or M_R is injective, then $\Delta(M_R) \subseteq \mathrm{Rad}(M_R)$.

2) If R_R is injective, then $\Delta(R_R) \subseteq \mathrm{Rad}(R)$.

There is a correspondence between ideals of endomorphism rings and bi-submodules of Hom.

Theorem VII.6.3. Let A and M be right R-modules, $S = \mathrm{End}(M_R)$, and $T = \mathrm{End}(A_R)$. Let $\mathcal{L}(S)$ denote the lattice of all two-sided ideals of S and $\mathcal{L}(H)$ the lattice of all S-T-submodules of H. Then

$$\mathrm{Mdl} : \mathcal{L}(S) \to \mathcal{L}(H) : \mathrm{Mdl}(I) = \{f \in H \mid f\, \mathrm{Hom}_R(M, A) \subseteq I\}$$

and

$$\mathrm{Idl} : \mathcal{L}(H) \to \mathcal{L}(S) : \mathrm{Idl}(K) = \textstyle\sum_{g \in \mathrm{Hom}_R(M, A)} Kg$$

are inclusion preserving maps with

$$\mathrm{Mdl}(\mathrm{Idl}(\mathrm{Mdl}(I))) = \mathrm{Mdl}(I) \text{ and } \mathrm{Idl}(\mathrm{Mdl}(\mathrm{Idl}(K))) = \mathrm{Idl}(K).$$

In Chapter VIII we deal with regularity in homomorphism groups of abelian groups.

The theory of abelian groups is a highly developed subject so that conclusive answers are possible using the available tools.

We wish to compute the maximum regular bi-submodule of $\mathrm{Hom}(A, M)$. A first step consists in checking when $\mathbb{Z}f$ is regular for a regular homomorphism $f \in \mathrm{Hom}(A.M)$.

Proposition VIII.2.9. Let A be a group and let $f \in \mathrm{Hom}(A, M)$. Then $\mathbb{Z}f = \{nf \mid n \in \mathbb{Z}\}$ is a regular subgroup of $\mathrm{Hom}(A, M)$ if and only if $\mathrm{Ker}(f)$ is a direct summand of A, $\mathrm{Im}(f)$ is a direct summand of M, and for all $n \in \mathbb{Z}$,

$$\mathrm{Im}(f) = n\,\mathrm{Im}(f) \oplus \mathrm{Im}(f)[n].$$

We can settle the torsion-free and torsion cases.

Theorem VIII.2.13. Let A and M be torsion-free abelian groups. Then $\mathrm{Reg}(A, M) = 0$ unless A and M are both divisible. If both A and M are divisible, then $\mathrm{Reg}(A, M) = \mathrm{Hom}(A, M)$.

Theorem VIII.2.14. Let A and M be p-primary groups.

1) Suppose that M is not reduced. Then $\mathrm{Reg}(A, M) = 0$.

2) Suppose that M is reduced. There are decompositions $A = A_1 \oplus A_2$ and $M = M_1 \oplus M_2$ such that A_1 and M_1 are elementary, and A_2 and M_2 have no direct summands of order p. Then

    i) $\mathrm{Reg}(A, M) = 0$ if $M_2 \neq 0$,

    ii) $\mathrm{Reg}(A, M) = 0$ if A_2 is not divisible,

    iii) $\mathrm{Reg}(A, M) = \mathrm{Hom}(A, M)$ if $M_2 = 0$ and A_2 is divisible, i.e., if A is the direct sum of an elementary group and a divisible group and M is elementary.

As usual the general case of a mixed group is much more difficult. If $f \in \mathrm{Reg}(A, M)$, then pf must be regular for every prime p. This fact alone has far-reaching consequences.

Theorem VIII.3.6. Let A and M be reduced abelian groups and assume that $\mathrm{Reg}(A, M) \neq 0$. Then there exists a non-void set of primes $\mathbb{P}(A, M)$ such that

$$\forall p \in \mathbb{P}(A, M), \ A = pA \oplus A[p], \quad A[p] \neq 0,$$

and

$$\forall p \in \mathbb{P}(A, M), \ M = M[p] \oplus M', \quad M[p] \neq 0.$$

The equations $A = pA \oplus A[p]$ clearly are very restrictive.

Corollary VIII.3.10. Suppose that A is a reduced abelian group such that, for every prime $p \in \mathbb{P}$, there is a decomposition $A = pA \oplus A[p]$. Then $A \cong G$ where G is a group with $\bigoplus_{p \in \mathbb{P}} A[p] \subseteq G \subseteq \prod_{p \in \mathbb{P}} A[p]$ such that $G / \bigoplus_{p \in \mathbb{P}} A[p]$ is divisible.

If $\mathrm{Reg}(A, M) \neq 0$ we are dealing with groups of the kind in Corollary 3.10 in the best of situations. These groups appear to be very concrete and accessible yet numerous studies have shown that they are very elusive. Let $\mathbf{t}(G)$ denote the maximal torsion subgroup of G.

For each $p \in \mathbb{P}$, let T_p be a zero or nonzero elementary p-group, set $P = \prod_{p \in \mathbb{P}} T_p$, $T = \bigoplus_{p \in \mathbb{P}} T_p = \mathbf{t}(P)$, and let $\mathbb{P}(T) = \{p \in \mathbb{P} \mid T_p \neq 0\}$. A group A is a $\mathcal{G}^*(T)$-**group** if $T \subseteq A \subseteq P$ such that A/T is (torsion-free) and $\mathbb{P}(T)$-divisible. A $\mathcal{G}^*(T)$-group A will be called **slim** if T_p is cyclic or zero for every $p \in \mathbb{P}$.

Theorem VIII.3.16. Suppose that $A \in \mathcal{G}^*(T_1)$ and $M \in \mathcal{G}^*(T_2)$. Then $\mathbf{t}(\mathrm{Hom}(A, M))$ is a regular bi-submodule of $\mathrm{Hom}(A, M)$.

There are $\mathcal{G}^*(T)$-groups A, M such that $\mathrm{Hom}_R(A, M)$ contains non-regular homomorphisms.

The situation is somewhat easier if $A = M$, which is the study of regularity in endomorphism rings of mixed groups. Despite numerous attempts ([29], [12], [30],

[13], [24]), it is not known which mixed abelian groups have regular endomorphism rings but there is one conclusive result.

Theorem VIII.4.2. ([13, Theorem 4.1]) Let A be a slim $\mathcal{G}^*(T)$-group such that A/T has finite rank and is divisible. Then $\text{End}(A)$ is regular.

The definition of regularity makes sense in any category and this is the topic of Chapter IX. In preadditive categories $\text{Reg}(A, M)$ exists as it does in module categories.

Theorem IX.1.3. Let $\mathcal{C}$ be a preadditive category, A, M objects of $\mathcal{C}$, $S = \mathcal{C}(M, M)$ and $T = \mathcal{C}(A, A)$. Then $\text{Reg}(A, M) = \{f \in \mathcal{C}(A, M) \mid SfT \text{ is regular}\}$ is the largest regular S-T-submodule of $\mathcal{C}(A, M)$.

Preadditive categories are looked at in some detail, in particular, kernels, cokernels, images and coimages of morphisms in a category are needed for further discussions of regularity.

Let A be an object in a category $\mathcal{C}$ and $e = e^2 \in \mathcal{C}(A, A)$. Then the **idempotent** e **splits in** $\mathcal{C}$ if there exists an object M and mappings $\iota \in \mathcal{C}(M, A)$, $\pi \in \mathcal{C}(A, M)$ such that $\iota\pi = e$ and $\pi\iota = 1_M$. We say that **idempotents split** in $\mathcal{C}$ if all idempotents in $\mathcal{C}$ are splitting. The splitting of idempotents means that every idempotent determines a direct decomposition. By $A \in_{\mathcal{C}} A_1 \oplus A_2$ we mean that there exists a set of structural maps (insertions and projections) that make A the biproduct of A_1 and A_2.

If idempotents split in a preadditive category $\mathcal{C}$, then we have the familiar decompositions associated with regular maps.

Theorem IX.2.7. Let $\mathcal{C}$ be a preadditive category in which idempotents split and suppose that $f \in \mathcal{C}(A, M)$ is regular and $fgf = f$ for $g \in \mathcal{C}(M, A)$. Then the following statements hold.

1) $e = fg \in \mathcal{C}(M, M)$ is an idempotent.

2) $d = gf \in \mathcal{C}(A, A)$ is an idempotent.

3) There are structural maps $\iota_{A_i} : A_i \to A$, and $\pi_{A_i} : A \to A_i$ such that $d = \iota_{A_1}\pi_{A_1}, \pi_{A_1}\iota_{A_1} = 1_{A_1}, 1_A - d = \iota_{A_2}\pi_{A_2}, \pi_{A_2}\iota_{A_2} = 1_{A_2}$, and $A \in_{\mathcal{C}} A_1 \oplus A_2$.

4) There are structural maps $\iota_{M_i} : M_i \to M$, and $\pi_{M_i} : M \to M_i$ such that $e = \iota_{M_1}\pi_{M_1}, \pi_{M_1}\iota_{M_1} = 1_{M_1}, 1_M - e = \iota_{M_2}\pi_{M_2}, \pi_{M_2}\iota_{M_2} = 1_{M_2}$, and $M \in_{\mathcal{C}} M_1 \oplus M_2$.

5) $\pi_{M_1}f\iota_{A_1} : A_1 \to M_1$ is an isomorphism with inverse $\pi_{A_1}g\iota_{M_1}$.

6) $\iota_{A_2} \in \text{Ker}(f)$ and $\iota_{M_2} \in \text{Ker}(fg)$.

7) $\iota_{M_1} \in \text{Im}(f)$ and $\iota_{A_1} \in \text{Im}(gf)$.

In the preadditive category $\mathcal{A}$ of torsion-free abelian groups of finite rank there is only one group whose endomorphism ring is a division ring, namely $\mathbb{Q}$, therefore regularity is not very interesting. Also there are highly non-unique direct decompositions in this category. To remedy this problem the **quasi-isomorphism category** $\mathbb{Q}\mathcal{A}$ was introduced by Bjarni Jonsson ([16], [17]). The objects of $\mathbb{Q}\mathcal{A}$ are the same as those of $\mathcal{A}$, namely all torsion-free abelian groups of finite rank, which is the same as all additive subgroups of finite-dimensional $\mathbb{Q}$-vector spaces. However, the morphism groups are now

$$\mathbb{Q}\operatorname{Hom}(A, M) = \{rf \mid r \in \mathbb{Q}, f \in \operatorname{Hom}(A, M)\} \subseteq \operatorname{Hom}(\mathbb{Q}A, \mathbb{Q}M)$$

where $\mathbb{Q}A$ and $\mathbb{Q}M$ are the vector subspaces spanned by A and M in the $\mathbb{Q}$-vector spaces in which they are contained as additive subgroups, and $\operatorname{Hom}(\mathbb{Q}A, \mathbb{Q}M)$ is just the groups of linear transformations. The category $\mathbb{Q}\mathcal{A}$ is a Krull-Schmidt category, i.e., every object is "uniquely" the direct sum of indecomposable objects. Also this category contains an abundance of groups whose endomorphism rings are division rings. The regularity picture in the category $\mathbb{Q}\mathcal{A}$ is transparent but there are no conclusive simple theorems because the indecomposable objects are too numerous, complex, and unknown.

Finally, we look at regularity in the category of (non-commutative) groups where semidirect products appear.

Theorem IX.5.2. Let G and H be groups and $f \in \operatorname{Hom}(G, H)$.

1) Assume that f is regular. Let $g \in \operatorname{Hom}(H, G)$ such that $fgf = f$. Then $e = gf \in \operatorname{End}(G)$ is an idempotent, $d = fg \in \operatorname{End}(H)$ is an idempotent and

$$G = \operatorname{Ker}(f) \rtimes \operatorname{Im}(e), \quad \operatorname{Im}(e) \cong \operatorname{Im}(f), \quad \text{and} \quad H = \operatorname{Ker}(d) \rtimes \operatorname{Im}(f).$$

2) Suppose that $G = \operatorname{Ker}(f) \rtimes K$ and $H = N \rtimes \operatorname{Im}(f)$. Then f is regular.

Chapter I

Notation and Background

1 Notation

Citations of numbered items belong to the chapter in which they appear, unless it is explicitly stated that the item is from a different chapter (e.g., Theorem II.2.3). When the citation contains two index numbers (e.g., Theorem 1.1), it refers to the theorem in general. When the citation contains three index numbers (e.g., Corollary II.3.5.2), it refers to Item number 2 in Corollary II.3.5.

In some statements $\forall x : \ldots$ the colon must be read as "it is true that", and in statements $\exists x : \ldots$ the colon must be read as "such that".

We use := to mean "by definition equal"; e.g., frequently used abbreviations are $H := \operatorname{Hom}_R(A, M)$, $S := \operatorname{End}(M_R)$, and $T := \operatorname{End}(A_R)$ where A and M are right R-modules.

We mostly use the following notation to define functions.

$$f : A \ni a \mapsto f(a) \in B.$$

In this way we avoid the awkward "where a is an element of A". For $X \subseteq B$, the term $f^{-1}(X) := \{a \in A \mid f(a) \in X\}$ denotes the preimage.

- $\mathbb{P}$ denotes the set of all prime numbers.

- Let $\mathbb{P}'$ be a set of primes. A natural number n is a $\mathbb{P}'$-**number** if the prime factors of n all belong to $\mathbb{P}'$.

- $\mathbb{N}_0 := \{0, 1, 2, \ldots\}$, $\mathbb{N} := \{1, 2, \ldots\}$ denote the set of natural numbers, one including 0 and the other excluding 0.

- $\mathbb{Q}$ is the field or group of rational numbers.

Let $A = B \oplus C$. Associated with such a decomposition we have the **projections**

$$\operatorname{pr}_B : A \to B \text{ along } C, \text{ and } \operatorname{pr}_C : A \to C \text{ along } B.$$

These projections can also be viewed as endomorphisms of A:

$$\text{pr}_B : A \ni a \mapsto \text{pr}_B(a) \in A, \quad \text{and} \quad \text{pr}_C : A \ni a \mapsto \text{pr}_C(a) \in A.$$

When viewed as endomorphisms we call the projections **projectors**.

Let M be a module and K a submodule of M. A submodule L of M is a **complement** of K in M if L is maximal with respect to the condition that $K \cap L = 0$. By Zorn's Lemma every submodule K has complements. A submodule N of M is a **supplement** of K in M if N is minimal with respect to the property that $K + N = M$. Supplements may or may not exist.

Let A_R and M_R be right R-modules. We apply maps of $\text{Hom}_R(A, M)$ on the left side of $a \in A$. If $\psi \in \text{Hom}_R(A, M)$, $s \in S := \text{End}(M_R)$, $t \in T := \text{End}(A_R)$, and $a \in A$, then

$$(s\psi t)(a) = (s \circ \psi \circ t)(a) = s(\psi(t(a)))$$

which shows that $\text{Hom}_R(A, M)$ is an S-T-bimodule. On the other hand, if $_RA$ and $_RM$ are left R-modules, then we apply maps of $\text{Hom}_R(A, M)$ on the right side of $a \in A$. If $\psi \in \text{Hom}_R(A, M)$, $s \in S := \text{End}(_RM)$, $t \in T := \text{End}(_RA)$, and $a \in A$, then

$$(a)(t\psi s) = (a)(t \circ \psi \circ s) = (((a)t)\psi)s$$

which shows that $\text{Hom}_R(A, M)$ is a T-S-bimodule.

2   Rings and Modules

The notation $K \subseteq^{\oplus} M$ means that K is a direct summand of the module M.

A submodule K of M_R is **small** or superfluous if whenever $N + K = M$ for some submodule N of M, then $N = M$. We write $K \subseteq^{\circ} M$ in this case.

A submodule L of M_R is **large** or essential if every nonzero submodule of M intersects L non-trivially. We write $L \subseteq^{*} M$ in this case.

The following is [20, Lemma 0.4].

Lemma 2.1. *Let $M = A \oplus B$ be a direct decomposition of R-modules. Denote by $\mathcal{C}$ the set of all direct complements of A in M. Then*

$$\text{Hom}(B, A) \ni \varphi \mapsto B_\varphi := \{\varphi(b) + b : b \in B\} \in \mathcal{C}$$

defines a bijective mapping and

$$\varphi + 1 : B \ni b \mapsto \varphi(b) + b \in B_\varphi$$

is an isomorphism.

Let M_R be an R-module and denote by $\text{Rad}(M)$ the **radical** of R. Recall that the following are equivalent descriptions of $\text{Rad}(M)$ ([18]).

$$
\begin{aligned}
\text{Rad}(M) &= \sum \{A \mid A \subseteq^{\circ} M\} = \bigcap \{B \mid B \text{ is maximal in } M\} \\
&= \bigcap \{\text{Ker}(\phi) \mid \phi \in \text{Hom}_R(M, N), N \text{ semisimple}\}.
\end{aligned}
$$

The radical $\mathrm{Rad}(R)$ of a ring R is $\mathrm{Rad}(R_R)$ which turns out to be the same as $\mathrm{Rad}(_RR)$.

Lemma 2.2. *Let R be a ring, A and M right R-modules, $S := \mathrm{End}(M_R)$, and $T := \mathrm{End}(A_R)$.*

1) *Let K be a non-void subset of R that is closed under right multiplication by elements of R. Then $K \subseteq \mathrm{Rad}(R)$ if and only if for all $a \in K$, $1 - a$ is an invertible element in R.*

2) *Let $f \in \mathrm{Hom}_R(A, M)$ and $g \in \mathrm{Hom}_R(M, A)$. Then $1_S - fg$ is invertible in S if and only if $1_T - gf$ is invertible in T.*

3)
$$\{f \in \mathrm{Hom}_R(A, M) \mid \forall g \in \mathrm{Hom}_R(M, A), gf \in \mathrm{Rad}(T)\}$$
$$= \{f \in \mathrm{Hom}_R(A, M) \mid \forall g \in \mathrm{Hom}_R(M, A), fg \in \mathrm{Rad}(S)\}.$$

Proof. 1) Assume first that $K \subseteq \mathrm{Rad}(R)$. Then, for every $a \in K$, $aR \subseteq^\circ R_R$, and hence $(1 - a)R + aR = R$ implies that $(1 - a)R = R$. So there is $s \in R$ such that $(1-a)s = 1$ and so $1 + as = s$. Since also $a(-s) \in K$, by specialization, there exists $t \in R$ with $(1 + as)t = st = 1$. It follows that $1 - a = (1 - a)st = ((1 - a)s)t = t$ and further $1 = st = s(1 - a)$. Hence s is the inverse of $1 - a$.

Conversely, assume that for every $a \in K$, $1 - a$ is an invertible element in R. Denote by $|K)$ the right ideal generated by K. We will show that $|K) \subseteq^\circ R_R$ which means that $|K) \subseteq \mathrm{Rad}(R)$. Assume that L is a right ideal of R with $R = L + |K)$. Then

$$1 = b + a_1 + \cdots + a_m, \quad b \in L, \ a_i \in K. \tag{1}$$

If $1 = b \in L$, then $L = R$ and we are done. Therefore we can assume that a_m with $m \geq 1$ really occurs. By assumption,

$$1 - a_m = b + a_1 + \cdots + a_{m-1}$$

is invertible, hence

$$1 = (b + a_1 + \cdots + a_{m-1})(1 - a_m)^{-1}. \tag{2}$$

Since $b(1 - a_m)^{-1} \in L$ and $a_i(1 - a_m)^{-1} \in A$ for $i = 1, \ldots, m - 1$, we have in (2) a representation analogous to (1) but with only $m - 1$ or no (for $m = 1$) summands from K. By induction we get that $1 \in L$ and $|K) \subseteq \mathrm{Rad}(R)$.

2) Suppose that $1_S - fg$ is invertible in S. It is easy to check that

$$(1_T - gf)^{-1} = 1_T + g(1_S - fg)^{-1}f.$$

Similarly, if $1_T - gf$ is invertible in T, then

$$(1_S - fg)^{-1} = 1_S + f(1_T - gf)^{-1}g.$$

3) Suppose that $f \in \text{Hom}_R(A, M)$ and for every $g \in \text{Hom}_R(M, A)$, $fg \in \text{Rad}(S)$. For all $s \in S$, $t \in T$, $g \in \text{Hom}_R(M, A)$, also $tgs \in \text{Hom}_R(M, A)$, and so $ftgs \in \text{Rad}(S)$. Hence the right ideal $ftgS$ is contained in $\text{Rad}(S)$. By 1.) it follows that $1_S - ftgs$ is invertible in S. Then, by 2.), $Tgsf$ is a left ideal contained in $\text{Rad}(T)$. Therefore $f \in \{f \in \text{Hom}_R(A, M) \mid \forall g \in \text{Hom}_R(M, A), gf \in \text{Rad}(T)\}$. The reverse containment follows in a similar fashion. $\square$

The following result is the well-known Dual Basis Lemma for projective modules ([18, 5.4.3]).

Lemma 2.3. *The following properties of a module P_R are equivalent.*

1) *P is projective.*

2) *Given a family $\{y_i \mid i \in I\}$ of generators of P, there exists a family of functionals $\{\varphi_i \in P^* := \text{Hom}_R(P, R) \mid i \in I\}$ such that for every $x \in P$, all but a finite number of the ring elements $\varphi_i(x)$, $i \in I$, are equal to 0 and $x = \sum_{i \in I} y_i \varphi_i(x)$.*

3) *There exist families $\{y_i \mid i \in I\}$ of generators of P and $\{\varphi_i \in P^* := \text{Hom}_R(P, R) \mid i \in I\}$ such that for every $x \in P$, we have $x = \sum_{i \in I} y_i \varphi_i(x)$ with $\varphi_i(x) = 0$ for all but a finite number of i.*

The following result on flat modules can be found in [18, 10.5.2] or [1, 19.18].

Proposition 2.4. *Let $_S P$ be a projective module over the ring S with 1, and let K be a submodule of P. Then P/K is flat if and only if for every finitely generated right ideal I of R, it is true that $K \cap IM = IK$.*

Lemma 2.5. *Let M be a module and N a fully invariant subgroup of M. If $M = K \oplus L$, then $N = (N \cap K) \oplus (N \cap L)$.*

Proof. Let pr_K and pr_L be the projectors belonging to the decomposition $M = K \oplus L$. It is clear that $N \cap K \oplus N \cap L \subseteq N$. To show equality let $x \in N$. Then $x = \text{pr}_K(x) + \text{pr}_L(x) \in N \cap K \oplus N \cap L$ where it is used that N is fully invariant in M. $\square$

Remark. We have an S-S-homomorphism

$$\rho_S : \text{Hom}_R(A, M) \otimes_T \text{Hom}_R(M, A) \to S = \text{End}(M_R) \text{ given by } \rho_S(f, g) = f \circ g,$$

and similarly, a T-T-homomorphism

$$\rho_T : \text{Hom}_R(M, A) \otimes_S \text{Hom}_R(A, M) \to T = \text{End}(A_R) \text{ given by } \rho_T(g, f) = g \circ f.$$

Proof. It is evident that the map $\rho_S : \text{Hom}_R(A, M) \times \text{Hom}_R(M, A) \to S$ satisfies $\rho_S(ft, g) = \rho_A(f, tg)$ where $t \in T$ and $\rho_S(f_1 + f_2, g) = \rho_S(f_1, g) + \rho_S(f_2, g)$ and $\rho_S(f, g_1 + g_2) = \rho_S(f, g_1) + \rho_S(f, g_2)$ and hence induces a group homomorphism on the tensor product. Also $\rho_S(s_1 f, g s_2) = s_1 \rho_S(f, g) s_2$. The argument for ρ_T is the same. $\square$

1) ρ_S is surjective if and only if there exist $f_1, \ldots, f_t \in \operatorname{Hom}_R(A, M)$ and $g_1, \ldots g_t \in \operatorname{Hom}_R(M, A)$ such that

$$f_1 g_1 + \cdots + f_t g_t = 1_M. \tag{3}$$

Proof. Suppose that ρ_S is surjective. Then there is an element in the tensor product that maps to 1_M and an element in the tensor product is of the form $f_1 \otimes g_1 + \cdots + f_t \otimes g_t$. Conversely, if (3) exists, then clearly ρ_S is surjective. □

2) If ρ_S is surjective, then S_S is finitely generated.

Proof. Suppose that ρ_S is surjective and (3) holds. Then $S = f_1 g_1 S + \cdots + f_t g_t S$, so S is generated by the $f_i g_i \in S$. □

If ρ_S and ρ_T are both surjective, then A and M must have interacting properties. If $A = R$, then $T = R$, and if $\rho_T = \rho_R$ is surjective, then M_R is projective. Another interesting case occurs when A is simple.

Question 2.6. Investigate situations where both ρ_S and ρ_T are surjective.

3   Abelian Groups

We now turn our attention to abelian groups. Notations and concepts on abelian groups are standard and can be found in [11].

- A **torsion group** is a group all of whose elements have finite order. A **torsion-free group** is a group all of whose nonzero elements have infinite order. A group is **mixed** if it may have elements both of finite and infinite order. The symbol $t(A)$ denotes the **maximal torsion subgroup** of A, i.e., the subgroup consisting of all torsion elements of A.

- A p-**primary group** or p-**group** is a torsion group all of whose elements have p-power order. A **primary group** is a group that is a p-group for some prime p.

- *Every torsion group T has a unique* **primary decomposition** $T = \bigoplus_{p \in \mathbb{P}} T_p$ *where T_p is p-primary.* The p-**components** T_p are fully invariant subgroups of T. For a general group G, we use $G_p := t(G)_p = \{x \in G \mid \exists n \in \mathbb{N} : p^n x = 0\}$.

- $\mathbb{Z}(n)$ denotes the cyclic group of order n.

- A group A is **divisible** if $nA = A$ for every positive integer n. A group A is p-**divisible** if $pA = A$. Let $\mathbb{P}'$ be a set of primes. A group D is $\mathbb{P}'$-**divisible** if $pD = D$ for every $p \in \mathbb{P}'$. *Every group contains a unique largest subgroup that is $\mathbb{P}'$-divisible. A group is divisible if and only if it is p-divisible for each $p \in \mathbb{P}$. The divisible abelian groups are exactly the injective abelian groups.*

- $\mathbb{Z}(p^\infty)$ denotes the indecomposable divisible p-group; it is isomorphic to the p-primary component of $\mathbb{Q}/\mathbb{Z}$, i.e., $\mathbb{Q}/\mathbb{Z} \cong \bigoplus_{p \in \mathbb{P}} \mathbb{Z}(p^\infty)$. It is known that $\operatorname{End}(\mathbb{Z}(p^\infty)) \cong \hat{\mathbb{Z}}_p$ where $\hat{\mathbb{Z}}_p$ is the ring of p-adic integers ([11, Volume I, page 181, Example 3]). Every divisible (= injective) abelian group is the direct sum of copies of the groups $\mathbb{Q}$ and $\mathbb{Z}(p^\infty)$, $p \in \mathbb{P}$.

- A group A is n-**bounded** if $nA = 0$ for some nonzero integer n.

- A p-group A is p-**elementary** if and only if it is a direct sum of groups of order p. Equivalently, A is p-elementary if and only if A is p-bounded. A group A is **elementary** if it is the direct sum of p-elementary groups. An abelian group A is semisimple as a $\mathbb{Z}$-module if and only if it is elementary.

- Every group A has a decomposition $A = B \oplus C$ such that C is divisible (= injective) and B is **reduced**, i.e., B contains no nonzero divisible subgroup.

- The p-**socle** of a group A is the subgroup $A[p] := \{x \in A : px = 0\}$. More generally, $A[n] := \{x \in A : nx = 0\}$. If A is a p-primary group, then $\bigcup_n A[p^n] = A$.

- A subgroup B of a group A is **pure** if $B \cap nA = nB$ for all integers n. *A bounded pure subgroup is a direct summand. The union of an ascending chain of pure subgroups is again pure.*

- A subgroup B of a group A is p-**pure** if $B \cap p^n A = p^n B$ for all positive integers n. For a set of primes $\mathbb{P}_0$ a subgroup B of A is $\mathbb{P}_0$-**pure** if B is p-pure for every $p \in \mathbb{P}_0$. *A subgroup B is pure in A if and only if it is p-pure for every prime p.*

Let A be a torsion-free abelian group. Then $A \cong \mathbb{Z} \otimes A \subseteq \mathbb{Q} \otimes A$ shows that A is embedded in a $\mathbb{Q}$-vector space. Hence, given a torsion-free abelian group A, it may always be assumed that A is an additive subgroup of a $\mathbb{Q}$-vector space V, and doing so the $\mathbb{Q}$-subspace $\mathbb{Q}A$ spanned by A is the smallest divisible subgroup of V containing A. It can be shown easily that $\mathbb{Q}A \cong \mathbb{Q} \otimes A$. This means that $\mathbb{Q}A$ is essentially unique and can, regardless of its containing vector space, always be thought of as $\mathbb{Q}A = \{ra \mid r \in \mathbb{Q}, a \in A\}$. The group $\mathbb{Q}A$ is the **divisible hull** of A.

Let A and B be torsion-free groups and $\mathbb{Q}A$, $\mathbb{Q}B$ their divisible hulls. Then

$$\mathrm{Hom}(\mathbb{Q}A, \mathbb{Q}B) = \mathrm{Hom}_{\mathbb{Z}}(\mathbb{Q}A, \mathbb{Q}B) = \mathrm{Hom}_{\mathbb{Q}}(\mathbb{Q}A, \mathbb{Q}B),$$

that is to say that the group homomorphisms on the divisible hulls are exactly the linear transformations on $\mathbb{Q}A$ to $\mathbb{Q}B$ as vector spaces.

Lemma 3.1. *Let A be an abelian group and $n \in \mathbb{Z}$. Then A has a maximal n-bounded direct summand.*

Proof. An n-bounded direct summand is an n-bounded pure subgroup and n-bounded pure subgroups are direct summands. It is well-known and easy to see that an ascending chain of pure subgroups is again pure. Also, an ascending chain of n-bounded subgroups is n-bounded. Therefore a straightforward application of Zorn's Lemma establishes that there exist maximal n-bounded pure subgroups. $\square$

In the context of regularity we will have to consider the following aspect of an abelian group.

Proposition 3.2. *Let A be any abelian group. Then for every $p \in \mathbb{P}$ there exists a direct decomposition*

$$A = A_1^p \oplus A_2^p$$

such that A_1^p is an elementary p-group and A_2^p has no direct summands of order p. Let $\pi_1^p : A \to A_1^p$ be the projection along A_2^p and define

$$\pi_1 : A \ni x \mapsto \pi_1(x) := \prod\nolimits_{p \in \mathbb{P}} \pi_1^p(x) \in \prod\nolimits_{p \in \mathbb{P}} A_1^p.$$

Then

$$\bigoplus\nolimits_{p \in \mathbb{P}} A_1^p \subseteq \pi_1(A) \subseteq \prod\nolimits_{p \in \mathbb{P}} A_1^p, \quad \mathrm{Ker}(\pi_1) = \bigcap\nolimits_{p \in \mathbb{P}} A_2^p,$$

and $\mathrm{Ker}(\pi_1)$ has no direct summands of prime order.

Let $\mathbb{P}(A) := \{p \in \mathbb{P} \mid A_1^p \neq 0\}$ and assume that A is reduced and

$$\forall p \in \mathbb{P}(A) : \mathrm{Hom}(A_2^p, A_1^p) = 0.$$

Then

$$\forall p \in \mathbb{P}(A) : A = A[p] \oplus pA,$$

$\mathrm{Ker}(\pi_1)$ *is the maximal* $\mathbb{P}(A)$*-divisible subgroup of* A *and* $A/\mathbf{t}(A)$ *is* $\mathbb{P}(A)$*-divisible.*

Proof. The decompositions exist by Lemma 3.1. The map π_1 is evidently well-defined, it acts as identity on $\bigoplus_{p \in \mathbb{P}} A_1^p$, and $\mathrm{Ker}(\pi_1)$ is as claimed.

Suppose that A is reduced, $A_1^p \neq 0$ and $\mathrm{Hom}(A_2^p, A_1^p) = 0$. Then $A_2^p = pA_2^p$ and $(A_2^p)_p$ is a pure subgroup of A_2^p, hence also p-divisible. But p-divisible p-groups are divisible and as A is reduced, $(A_2^p)_p = 0$. It follows from $A = A_1^p \oplus A_2^p$ that $A[p] = A_1^p$ and $pA = pA_2^p = A_2^p$. $\qquad\qquad\square$

Remark. In the decomposition $A = A_1^p \oplus A_2^p$ of Proposition 3.2, the summand A_1^p is unique if and only if $A_2^p = pA_2^p$ and the summand A_1^p is unique if and only if $A_2^p[p] = 0$ (Lemma 2.1).

The following examples show that in general little can be said about the quotient $\pi_1(A)/\mathbf{t}(\pi_1(A))$ and $\mathrm{Ker}(\pi_1)$.

Example 3.3. Let X be a torsion-free group of rank $\leq 2^{\aleph_0}$. Then there is a group A such that $\pi_1(A)/\mathbf{t}(\pi_1(A)) \cong X$. In fact, $\forall p \in \mathbb{P}$, let T_p be cyclic of order p, let $T := \bigoplus_{p \in \mathbb{P}} T_p$ and $P := \prod_{p \in \mathbb{P}} T_p$. Then P/T is torsion-free divisible (a $\mathbb{Q}$-vector space) of rank $2^{\aleph_0}$. Without loss of generality $X \subseteq P/T$ and let A be the group with $T \subseteq A \subseteq P$ and $A/T = X$. Then $\pi_1(\mathbf{t}(A)) = \mathbf{t}(\pi_1(A)) = T$ and $\pi_1(A) = A$.

Example 3.4. Let T_p be cyclic of order p, let $T := \bigoplus_{p \in \mathbb{P}} T_p$ and let K be any group that has no direct summand of prime order. Let $A := T \oplus K$, $A_1^p := T_p$ and $A_2^p := K \oplus \bigoplus_{q \neq p} T_q$. Then $\pi_1(A) = T$ and $\mathrm{Ker}(\pi_1) = K$.

Let A be an abelian group and B a subgroup of A. A subgroup H of A containing B is called a **pure hull** of B in A if H is pure in A and if $B \subseteq H' \subseteq H$ and H' is pure in A, then $H' = H$. In general the questions of existence and uniqueness properties of pure hulls are difficult and not completely resolved. However, in special cases, which includes the case of a torsion-free group A, there is an easy answer.

Lemma 3.5. *Let* A *be an abelian group,* T *the maximal torsion subgroup of* A *and* B *a subgroup of* A *containing* T*. Then there exists a unique smallest pure subgroup* B_*^A *of* A *containing* B*. Furthermore,* A/B_*^A *is torsion-free and* B_*^A/B *is a torsion group. We call* B_*^A *the* **purification** *of* B *in* A*.*

Proof. Let B_*^A be the unique subgroup of A such that $B_*^A/B = \mathbf{t}(A/B)$. The claims are easily verified. $\qquad\qquad\square$

Lemma 3.6. *Let* $D \subseteq N$ *be abelian groups,* m *a positive integer and assume that* $mD = D$ *and* $m(N/D) = N/D$*. Then* $mN = N$*.*

Proof.

$$\frac{N}{D} \overset{\text{hyp.}}{=} m\frac{N}{D} = \frac{mN + D}{D} \overset{\text{hyp.}}{=} \frac{mN + mD}{D} = \frac{mN}{D}$$

so $mN = N$.   □

Proposition 3.7. *Let G be an Abelian group and H a subgroup of G. Then H is a small subgroup of G if and only if*

1) $H \subseteq \bigcap_{p \in \mathbb{P}} pG$,

2) H *has no nonzero divisible quotient group.*

Proof. (a) Suppose that H is small in G. If $pG = G$, then trivially $H \subseteq pG$. Suppose that $pG \neq G$. Then G/pG is a vector space over the prime field $\mathbb{Z}/p\mathbb{Z}$ and there is a family $\mathcal{K}$ of maximal subgroups of G containing pG whose intersection is pG. Let $K \in \mathcal{K}$. If H is not contained in K, then $H + K = G$ and $K \neq G$. As H is small we conclude that $H \subseteq K$ and so $H \subseteq \bigcap(K) = pG$. This, being true for every p, establishes 1).

To prove 2) we assume to the contrary that there is a subgroup L of H such that H/L is nonzero divisible (= injective). Then $G/L = H/L \oplus K/L$ for some subgroup K of G and $K \neq G$ since $H/L \neq 0$. Now $G = H + K$ and $G \neq K$, contrary to the smallness of H.

(b) Suppose now that 1.) and 2.) hold and that $G = H + K$ for some subgroup K of G. By 1.), $G = pG + K$ for any prime p, and hence $p(G/K) = (pG + K)/K = G/K$ for every prime p which says that G/K is divisible. Therefore $H/(H \cap K) \cong (H + K)/K = G/K$ is divisible, and by assumption 2) $H/(H \cap K) = 0$, so $H \subseteq K$ and $G = K$ showing that H is small.   □

It is much easier to describe the large subgroups.

Proposition 3.8. *A subgroup L of a group G is large in G if and only if $\mathrm{Soc}(G) = \sum_{p \in \mathbb{P}} G[p] \subseteq L$ and G/L is torsion.*

Proof. Suppose that L is large in G. If x is an element of prime order, then $x\mathbb{Z} \cap L \neq 0$ which says that $\mathrm{Soc}(G) \subseteq L$. Let $0 \neq x \in G$ be arbitrary. Then $x\mathbb{Z} \cap L \neq 0$, so G/L is torsion. Conversely, Suppose L is a subgroup containing $\mathrm{Soc}(G)$ and such that G/L is torsion. Let $x \in G$. As G/L is torsion there is a least positive integer n such that $xn \in L$. If $xn \neq 0$ we have the desired nonzero intersection of $x\mathbb{Z}$ and L. So suppose that $xn = 0$ and $n \neq 1$. Then, for a prime p dividing n, the element $x(n/p)$ is not in L by minimality of n and $p(x(n/p)) = 0$ which means that $x(n/p) \in \mathrm{Soc}(G) \subseteq L$, a contradiction.   □

Chapter II

Regular Homomorphisms

1   Definition and Characterization

Let R be a ring with $1 \in R$ and denote by Mod-R the category of all unitary right R-modules. For arbitrary $A, M \in$ Mod-R, let

$$H := \operatorname{Hom}_R(A, M), \quad S := \operatorname{End}(M_R), \quad T := \operatorname{End}(A_R).$$

Then H is an S-T-bimodule.

Definition 1.1. Let $f \in \operatorname{Hom}_R(A, M)$. Then f is called **regular** if there exists $g \in \operatorname{Hom}_R(M, A)$ such that

$$fgf = f. \tag{1}$$

If so, g is called a **quasi-inverse** of f (or **r-inverse** of f). A subset X of $\operatorname{Hom}_R(A, M)$ is **regular** if every element of X is regular.

The definition of regular homomorphism used here includes the classical cases of regularity in a ring and in a module. Indeed the isomorphism

$$\operatorname{Hom}_R(R, M) \ni f \mapsto f(1) \in M \tag{2}$$

brings us to M. Similarly, the isomorphism $\operatorname{Hom}_R(R, R) \cong R$ brings us to the usual definition of regularity in a ring.

An invertible element f is regular because $ff^{-1}f = f$. This means that "regular" is a generalization of invertible and motivates the term "quasi-inverse". We will come back to quasi-inverses later.

It follows from (1) that

$$f(gfg)f = f, \qquad (gfg)f(gfg) = gfg.$$

This shows that the quasi-inverse of f is not uniquely determined by f; g and gfg are both quasi-inverses of f. We also see that f is a quasi-inverse of the regular element gfg.

If (1) is multiplied by g on the left and also on the right, then we obtain two idempotents

$$e := fg = e^2 \in S, \quad d := gf = d^2 \in T. \tag{3}$$

By (1) and the definitions of e and d, it follows further that

$$ef = fgf = f = fd. \tag{4}$$

If $f \neq 0$, then also $e \neq 0$ and $d \neq 0$. Since e and d are idempotents, also $1 - e$ and $1 - d$ are idempotents and

$$M = e(M) \oplus (1-e)(M), \quad A = d(A) \oplus (1-d)(A). \tag{5}$$

The following theorem characterizes regular maps.

Theorem 1.2. *If $f \in \mathrm{Hom}_R(A, M)$ is regular and $fgf = f$, then*

$$A = \mathrm{Ker}(f) \oplus \mathrm{Im}(gf), \quad \mathrm{Im}(gf) \cong \mathrm{Im}(f) \cong \mathrm{Im}(fg), \tag{6}$$

and

$$M = \mathrm{Im}(f) \oplus \mathrm{Ker}(fg), \quad \mathrm{Ker}(fg) = \mathrm{Im}(1 - fg). \tag{7}$$

Furthermore, the mapping

$$f_0 : \mathrm{Im}(gf) \ni gf(a) \mapsto fgf(a) = f(a) \in \mathrm{Im}(f)$$

is an isomorphism with inverse isomorphism

$$g_0 : \mathrm{Im}(f) \ni f(a) \mapsto gf(a) \in \mathrm{Im}(gf).$$

Proof. Set $e = fg$ and $d = gf$. We have the decompositions (5). We show first that $(1 - d)(A) = \mathrm{Ker}(f)$. Indeed, by (4)

$$f(1 - d) = f - fd = 0,$$

so

$$f(1 - d)(A) = 0,$$

hence $(1 - d)(A) \subseteq \mathrm{Ker}(f)$. Now let $a \in \mathrm{Ker}(f)$. Then

$$(1 - d)(a) = a - gf(a) = a,$$

hence $\mathrm{Ker}(f) \subseteq (1 - d)(A)$. Together we have $\mathrm{Ker}(f) = (1 - d)(A)$. The complementary summand is $d(A) = gf(A)$ and $gf(A) \cong A/\mathrm{Ker}(f) \cong \mathrm{Im}(f)$.

Using (4) again, we get the chain

$$\mathrm{Im}(f) = f(A) = ef(A) \subseteq e(M) = fg(M) \subseteq f(A),$$

hence $f(A) = e(M) = fg(M)$ which is a direct summand of M. The complementary summand is $(1 - e)(A) = \mathrm{Ker}(e) = \mathrm{Ker}(fg)$.

Since $A = gf(A) \oplus \mathrm{Ker}(f)$, we see that f_0 is a monomorphism and f_0 is evidently surjective.

Let $x \in \mathrm{Im}(gf)$, say $x = gf(a)$. Then $g_0 f_0(x) = g_0 f_0(gf(a)) = gfgf(a) = gf(a) = x$ so $g_0 f_0 = 1$. Let $x \in \mathrm{Im}(f)$, say $x = f(a)$. Then $f_0 g_0(x) = fgf(a) = f(a) = x$, so $f_0 g_0 = 1$. $\qquad\square$

We come to an important characterization of regularity.

Corollary 1.3 (Characterization of Regularity). *Let $f \in \mathrm{Hom}_R(A, M)$. Then f is regular if and only if $\mathrm{Ker}(f) \subseteq^\oplus A$ and $\mathrm{Im}(f) \subseteq^\oplus M$.*

Proof. Suppose f is regular. Apply Theorem 1.2.

Conversely, assume that

$$A = \mathrm{Ker}(f) \oplus A_0 \quad \text{and} \quad M = \mathrm{Im}(f) \oplus M_0, \tag{8}$$

and let $\pi : M \to \mathrm{Im}(f)$ be the projection of M onto $\mathrm{Im}(f)$ along M_0, and let $\iota : A_0 \to A$ be the inclusion map. Define

$$f_0 : A_0 \ni a \mapsto f(a) \in \mathrm{Im}(f).$$

By (8) f_0 obviously is an isomorphism. Let $g := \iota f_0^{-1} \pi$. We will show that $fgf = f$. Let $a \in A$. Then, by (8), we have $a = u + a_0$, where $u \in \mathrm{Ker}(f)$ and $a_0 \in A_0$. It follows that

$$f(a) = f(a_0) \quad \text{and} \quad fgf(a) = fgf(a_0) = ff_0^{-1}f(a_0) = f(a_0) = f(a),$$

hence $fgf = f$. $\qquad\square$

We remark that Corollary 1.3 already establishes that groups and rings of matrices over fields are regular since matrices are linear transformations and kernels and images of linear transformations are direct summands.

By Corollary 1.5 it is easy to exhibit modules A and M such that $\mathrm{Hom}_R(A, M)$ contains no nonzero regular homomorphisms. On the other hand, there are modules A and M such that every element in $\mathrm{Hom}_R(A, M)$ is regular.

Corollary 1.4. *In the following three cases all elements of $\mathrm{Hom}_R(A, M)$ are regular.*

1) *A and M are both semisimple.*

2) *A is semisimple and injective and M is arbitrary.*

3) *M is semisimple and projective and A is arbitrary.*

Proof. 1) Clear by Corollary 1.3.

2) Let $0 \neq f \in H$. Then $\mathrm{Ker}(f)$ is a submodule of A, and, A being semisimple, a direct summand of A: $A = \mathrm{Ker}(f) \oplus A_0$. Since A is injective, A_0 is also injective, and so is $\mathrm{Im}(f) \cong A/\mathrm{Ker}(f) \cong A_0$. Thus $\mathrm{Im}(f) \subseteq^\oplus M$. By Corollary 1.3 it follows that f is regular.

3) This case is dual to 2.). Since $\mathrm{Im}(f)$ is a submodule of M and M is semisimple, $\mathrm{Im}(f)$ is a direct summand of M and then also projective. Hence $A/\mathrm{Ker}(f) \cong \mathrm{Im}(f)$ is projective and this means that $\mathrm{Ker}(f)$ is a direct summand of A. Again by Corollary 1.3, f is regular.   □

Corollary 1.5. $\mathrm{Hom}_R(A, M)$ *contains a regular map* $\neq 0$ *if and only if there exist nonzero direct summands of A and of M that are isomorphic. Thus* $\mathrm{Hom}_R(A, M)$ *contains nonzero regular maps if and only if* $\mathrm{Hom}_R(M, A)$ *contains nonzero regular maps.*

Proof. Let $f \neq 0$ be regular. The claim follows from Theorem 1.3 because $A_0 \cong A/\mathrm{Ker}(f) \cong \mathrm{Im}(f) \neq 0$.

The proof of the converse is also easy. We include it for the convenience of the reader. Suppose that

$$A = A_1 \oplus A_0, \quad M = M_1 \oplus M_0, \quad A_0 \cong M_1 \neq 0.$$

Let $f_0 : A_0 \to M_1$ be an isomorphism. Let

- π_0 be the projection of A onto A_0 along A_1,

- π_1 the projection of M onto M_1 along M_0,

- ι_0 the inclusion of A_0 in A, and

- ι_1 the inclusion of M_1 in M.

Define

$$f := \iota_1 f_0 \pi_0, \quad g := \iota_0 f_0^{-1} \pi_1.$$

It is now easy to see that $f \neq 0$ and that $fgf = f$.   □

The proof shows that f_0 is the restriction of the regular map f.

The case of regular maps that are mutual quasi-inverses is particularly nice.

Proposition 1.6. *The following statements hold.*

1) *Let $f \in \mathrm{Hom}_R(A, M)$ be regular. Then there exists a quasi-inverse g of f that is also regular and has quasi-inverse f, i.e., in addition to $fgf = f$ we have that $gfg = g$.*

2) *Suppose that $f \in \mathrm{Hom}_R(A, M)$ and $g \in \mathrm{Hom}_R(M, A)$ are such that $fgf = f$ and $gfg = g$. Then*

$$A = \mathrm{Ker}(f) \oplus \mathrm{Im}(g), \quad and \quad M = \mathrm{Im}(f) \oplus \mathrm{Ker}(g).$$

3) *Conversely, suppose that $f \in \mathrm{Hom}_R(A, M)$ and $g \in \mathrm{Hom}_R(M, A)$, $A = \mathrm{Ker}(f) \oplus \mathrm{Im}(g)$ and $M = \mathrm{Im}(f) \oplus \mathrm{Ker}(g)$. Then there is $\alpha \in \mathrm{Aut}(\mathrm{Im}(f))$ such that $\alpha f \in \mathrm{Hom}_R(A, M)$ and g are both regular and mutual quasi-inverses.*

4) *Suppose that $f \in \operatorname{Hom}_R(A, M)$. Let*

$$F := \{h \in \operatorname{Hom}_R(M, A) \mid fhf = f, hfh = h\}.$$

Then F is in bijective correspondence with $\operatorname{Hom}(M/\operatorname{Im}(f), \operatorname{Im}(f))$.

Proof. 1) Recall that $fgf = f$ implies

$$gfg\, f\, gfg = gfg, \quad f\, gfg\, f = f$$

saying that f is the quasi-inverse of the regular homomorphism gfg and f also has the quasi-inverse gfg.

2) We have decompositions $A = \operatorname{Ker}(f) \oplus \operatorname{Im}(gf)$ and similarly $M = \operatorname{Ker}(g) \oplus \operatorname{Im}(fg)$. We will show that $\operatorname{Im}(gf) = \operatorname{Im}(g)$ and $\operatorname{Im}(fg) = \operatorname{Im}(f)$. Obviously $g(M) \supseteq gf(A) \supseteq gfg(M) = g(M)$ which establishes that $\operatorname{Im}(gf) = \operatorname{Im} g$. By symmetry $\operatorname{Im}(fg) = \operatorname{Im}(f)$.

3) As $A = \operatorname{Ker}(f) \oplus \operatorname{Im}(g)$ and $M = \operatorname{Im}(f) \oplus \operatorname{Ker}(g)$, the map f induces an isomorphism $f_0 : \operatorname{Im}(g) \to \operatorname{Im}(f)$ and g induces an isomorphism $g_0 : \operatorname{Im}(f) \to \operatorname{Im}(g)$. Let $\alpha := (f_0 g_0)^{-1} \in \operatorname{Aut}(\operatorname{Im}(f))$. Then

$$(\alpha f)g(\alpha f) = \alpha f g g_0^{-1} f_0^{-1} f = \alpha f, \quad \text{and} \quad g(\alpha f)g = g g_0^{-1} f_0^{-1} fg = g.$$

4) By 2), for any $h \in F$, we have $M = \operatorname{Im}(f) \oplus \operatorname{Ker}(h)$. The assignment $h \mapsto \operatorname{Ker}(h)$ is clearly an injective assignment of h to a direct complement of $\operatorname{Im}(f)$ in M. To show that the assignment is surjective, suppose that $M = \operatorname{Im}(f) \oplus C$ for some C. We define $h : M \to A$ to act as 0 on C and to agree with f^{-1} on $\operatorname{Im}(f)$. It is routine to check that $fhf = f$ and $hfh = h$. We have shown that F is in bijective correspondence with the direct complements of $\operatorname{Im}(f)$ in M and by Lemma I.2.1 the set of complements in turn is in bijective correspondence with $\operatorname{Hom}(\operatorname{Ker}(g), \operatorname{Im}(f))$ and $\operatorname{Ker}(g) \cong M/\operatorname{Im}(f)$. $\square$

As a matter of curiosity we consider two special cases.

Remark. If f is regular and injective and $fgf = f$, then $gf = 1_A$, $M = \operatorname{Im}(f) \oplus \operatorname{Ker}(g)$ and $\operatorname{Im}(f) \cong A = \operatorname{Im}(g)$.

If f is regular and surjective and $fgf = f$, then $fg = 1_M$, $A = \operatorname{Ker}(f) \oplus \operatorname{Im}(g)$ and $\operatorname{Im}(f) = M \cong \operatorname{Im}(g)$.

Question 1.7. Describe all pairs A, M such that all elements in H are regular.

2 Partially Invertible Homomorphisms and Quasi-Inverses

We have already seen that idempotents are very important for the study of regularity. An idempotent e is regular and is its own quasi-inverse since $e^3 = e$. If

$f \in \mathrm{Hom}_R(A, M)$ is regular and $fgf = f$ for $g \in \mathrm{Hom}_R(M, A)$, then we have the idempotents

$$e := fg, \quad d := gf$$

and f and g are factors of these idempotents. Conversely, a factor of an idempotent is the quasi-inverse of some regular map as we will see next.

Theorem 2.1. *The following statements are equivalent for* $g \in \mathrm{Hom}_R(M, A)$.

1) *There exists* $h \in \mathrm{Hom}_R(A, M)$ *such that* $e := hg = e^2 \neq 0$.

2) *There exists* $k \in \mathrm{Hom}_R(A, M)$ *such that* $d := gk = d^2 \neq 0$.

3) *There exists* $f \in \mathrm{Hom}_R(A, M)$ *such that* $fgf = f \neq 0$.

4) *There exist* $0 \neq M_0 \subseteq^{\oplus} M$, *and* $A_0 \subseteq^{\oplus} A$ *such that*

$$g_0 : M_0 \ni x \mapsto g(x) \in A_0 \quad \text{is an isomorphism.}$$

Proof. 1) $\Rightarrow$ 2): Suppose that $e = hg = e^2 \neq 0$. Then for $k := eh$ and $d := gk = geh$ it follows that

$$d^2 = gehgeh = ge^3h = geh = d$$

and

$$hdg = hgkg = hgehg = e^3 = e \neq 0,$$

hence $d \neq 0$.

    2) $\Rightarrow$ 1): Similar to 1) $\Rightarrow$ 2). Now we have

$$(kdg)^2 = kdgkdg = kd^3g = kdg.$$

    1) $\Rightarrow$ 3): Let $f := eh$. Then

$$fgf = ehgeh = e^3h = eh = f$$

and

$$ehg = e^2 = e \neq 0, \quad \text{hence } f = eh \neq 0.$$

    3) $\Rightarrow$ 1): Since $fgf = f \neq 0$, it follows that $e = fg \neq 0$.

    3) $\Rightarrow$ 4): This follows from Theorem 1.2 with $M_0 = e(M)$ and $A_0 = d(A)$.

    4) $\Rightarrow$ 3): We choose f such that $fgf = f \neq 0$. By assumption we have

$$M = M_0 \oplus M_1, \quad A = A_0 \oplus A_1,$$

and we have the isomorphism

$$g_0 : M_0 \ni x \mapsto g(x) \in A_0, \ g_0 \neq 0.$$

Now define f as

$$f : A_0 \to M_0 : f = g_0^{-1}, \quad f : A_1 \to 0.$$

Then, for $x \in A$, $x = a_0 + a_1$ where $a_0 \in A_0$, $a_1 \in A_1$, hence

$$(fgf)(x) = fgf(a_0 + a_1) = fgf(a_1) = f(a_0) = f(x)$$

hence $fgf = f$. □

Definition 2.2. We assume that the conditions of Theorem 2.1 are satisfied. Then

1) g is called **partially invertible**;

2) the **total** of A to M is defined to be

$$\mathrm{Tot}(A, M) := \{f \in \mathrm{Hom}_R(A, M) \mid f \text{ is not partially invertible.}\}.$$

The monograph ([20]) contains an extensive study of the total $\mathrm{Tot}(A, M)$ of two R-modules. The total is in general not additively closed but it has the following multiplicative closure property.

Lemma 2.3 ([20, Chapter II, Corollary 1.10]). *If A, M, A', M' are right R-modules and $f \in \mathrm{Hom}_R(A', A)$, $g \in \mathrm{Hom}_R(M, M')$, then $f \mathrm{Tot}(A, M)g \subseteq \mathrm{Tot}(A', M')$.*

One of the major questions surrounding the total is to characterize the pairs A, M such that $\mathrm{Tot}(A, M)$ is additively closed. The total is essential if M is a module with LE-decomposition, i.e., $M = \bigoplus_{i \in I} M_i$ where for each i, $\mathrm{End}(M_i)$ is a local ring. In this case $\mathrm{Tot}(\mathrm{End}(M))$ is an ideal in $\mathrm{End}(M)$ and ([20, Chapter IV, Corollary 2.4]) *the quotient ring $\mathrm{End}(M)/\mathrm{Tot}(\mathrm{End}(M))$ is isomorphic to a product of endomorphism rings of vector spaces over division rings.*

By (3) every regular map is partially invertible. Thus we have the containments

$$\{f \in H \mid f \text{ is regular}\} \subseteq \{f \in H \mid f \text{ is partially invertible}\}$$

and

$$\mathrm{Tot}(A, M) \subseteq \{f \in \mathrm{Hom}_R(A, M) \mid f \text{ is NOT regular}\}.$$

The following example shows, unsurprisingly, that a partially invertible element need not be regular.

Example 2.4. Let p be a prime, and

$$A = \mathbb{Z}a \oplus \mathbb{Z}b, \quad M = \mathbb{Z}m \oplus \mathbb{Z}n$$

where a, b, m have order p and n has order p^2. Define

$$f : A \to M : f(a) = m, f(b) = pn, \quad g : M \to A : g(m) = a, g(n) = 0.$$

These are well-defined homomorphisms with the following properties.

1) $fg(m) = m$, $fg(n) = 0$, $gf(a) = a$, $gf(b) = 0$.

2) $e := fg$ is idempotent and is the projector of M onto $\mathbb{Z}m$ along $\mathbb{Z}n$.

3) $d := gf$ is idempotent and is the projector of A onto $\mathbb{Z}a$ along $\mathbb{Z}b$.

4) $gfg = g$.

5) $f_0 : \mathbb{Z}a \ni x \mapsto f(x) \in \mathbb{Z}m$ is an isomorphism.

6) f is partially invertible by 2) or 3) but not regular as $\mathrm{Im}(f) = \mathbb{Z}m \oplus \mathbb{Z}pn$ is not a direct summand of M.

The difference between a partially invertible and a regular map can be elucidated in general.

Remark. Suppose that $f \in \mathrm{Hom}_R(A, M)$, $g \in \mathrm{Hom}_R(M, A)$ and $gf = e = e^2 \in \mathrm{End}(A_R)$. Then $d := fgfg$ is an idempotent in $\mathrm{End}(M_R)$, and we have direct decompositions

$$A = eA \oplus (1 - e)A, \qquad M = dM \oplus (1 - d)M.$$

Now $dM \subseteq \mathrm{Im}(f)$ and dM is a direct summand of M but this does not necessarily mean that $\mathrm{Im}(f)$ is a direct summand of M, and the latter would have to happen if f were regular.

Remark. We have seen that $\{f \in H \mid f \text{ is partially invertible}\} = \{f \in H \mid f \text{ is a quasi-inverse}\}$. Does this set have any algebraic properties?

Let $A = \langle a \rangle$ be a cyclic group of order p and let $M = \langle m \rangle$ be a cyclic group of order p^2. Then $\mathrm{Hom}(A, M)$ contains no partially invertible element, so $\{f \in H \mid f \text{ is partially invertible}\} = \{f \in H \mid f \text{ is a quasi-inverse}\} = \emptyset$.

When does the regular map f have a unique quasi-inverse? This question is settled next.

Theorem 2.5. *Let $f \in \mathrm{Hom}_R(A, M)$. There is a bijective correspondence between quasi-inverses of f and $\mathrm{Hom}_R(M/\mathrm{Im}(f), A)$. Consequently, the regular homomorphism f has a unique quasi-inverse if and only if $\mathrm{Hom}_R(M/\mathrm{Im}(f), A) = 0$.*

Proof. We see from Theorem 1.2 that for any quasi-inverse g of f, the restriction of g to $\mathrm{Im}(f)$ is the same map, namely f_0^{-1}. Another quasi-inverse g' can therefore only differ from g on $\mathrm{Ker}(fg)$, and how it acts on $\mathrm{Ker}(fg)$ is without consequence as we can see from

$$f = fg'f \quad \text{if and only if} \quad f(A) = fg'f(A) = fg'(f(A))$$

where only the action of g' on $f(A)$ enters.

Now fix a quasi-inverse g. We have the decomposition $M = \mathrm{Im}(f) \oplus \mathrm{Ker}(fg)$. Let $h : M \to A$ be any other quasi-inverse of f. Then $h \restriction_{\mathrm{Ker}(fg)} \in \mathrm{Hom}_R(\mathrm{Ker}(fg), A)$ and the assignment $h \mapsto h \restriction_{\mathrm{Ker}(fg)}$ is injective since $h = f^{-1}$ on $\mathrm{Im}(f)$. The preceding remark establishes that the assignment is also surjective. Since $\mathrm{Ker}(fg) \cong M/\mathrm{Im}(f)$, the first claim is proved and the second is a trivial consequence of the first. $\qquad \square$

It turns out that the quasi-inverse of $f \in \mathrm{Hom}_R(A, M)$ is unique if $\mathrm{End}(M_R)$ is commutative.

Proposition 2.6. *Suppose that $f \in \mathrm{Hom}_R(A, M)$ is regular and $\mathrm{End}(M_R)$ is commutative. Then f has a unique quasi-inverse.*

Proof. Let g_1 and g_2 be quasi-inverses of f. We have the decompositions $M = \mathrm{Im}(f) \oplus \mathrm{Ker}(fg_1) = \mathrm{Im}(f) \oplus \mathrm{Ker}(fg_2)$. The direct summands of M are endomorphic images of M and hence fully invariant because $\mathrm{End}(M_R)$ is commutative. Therefore $\mathrm{Hom}_R(\mathrm{Ker}(fg_1), \mathrm{Im}(f)) = 0$ and this means that the complementary summand $\mathrm{Ker}(fg_1)$ of $\mathrm{Im}(f)$ is unique, so $\mathrm{Ker}(fg_1) = \mathrm{Ker}(fg_2)$. Thus both g_1 and g_2 map $\mathrm{Ker}(fg_1) = \mathrm{Ker}(fg_2)$ to 0 and on $\mathrm{Im}(f)$ they both agree with f^{-1}. It follows that $g_1 = g_2$. $\qquad\square$

Corollary 2.7. *Let $f \in \mathrm{Hom}_R(A, M)$ be regular with quasi-inverse $g \in \mathrm{Hom}_R(M, A)$ and suppose that $\mathrm{End}(A_R)$ is commutative. Then f is the unique quasi-inverse of gfg.*

Proof. We have seen that $(gfg)f(gfg) = gfg$, so gfg is regular with quasi-inverse f that is unique by Proposition 2.6. $\qquad\square$

3   Regular Homomorphisms Generate Projective Direct Summands

In this short section we establish interesting properties of regular homomorphisms.

Theorem 3.1. *Let $S := \mathrm{End}(M_R)$, $T := \mathrm{End}(A_R)$ and assume that $0 \neq f \in \mathrm{Hom}_R(A, M)$ is regular. Then the following statements hold.*

1) *Sf is a nonzero S-projective direct summand of $_S\mathrm{Hom}_R(A, M)$ that is isomorphic to a cyclic left ideal of S, and this cyclic left ideal is a direct summand of S.*

2) *fT is a nonzero T-projective direct summand of $\mathrm{Hom}_R(A, M)_T$ that is isomorphic to a cyclic right ideal of T and this cyclic right ideal of T is a direct summand of T.*

More precisely, let $g \in \mathrm{Hom}_R(M, A)$ be a quasi-inverse of f and set $H := \mathrm{Hom}_R(A, M)$. Then:

1) *$S = Sfg \oplus S(1 - fg)$, $H = Sf \oplus H(1 - gf)$, and $Sf \cong Sfg$.*

2) *$T = gfT \oplus (1 - gf)T$, $H = fT \oplus (1 - fg)H$, and $fT \cong gfT$.*

Proof. 1) Since $0 \neq f \in H$ is regular and $fgf = f$ for $g \in \mathrm{Hom}_R(M, A)$, the mapping $e := fg$ is a nonzero idempotent in S, and

$$S = Se \oplus S(1 - e)$$

where $Se \neq 0$ and Se is a projective left ideal of S. We claim that the S-homomorphism

$$\sigma : Se \ni se \to sef = sf \in Sf$$

is an isomorphism, showing that Sf is also projective. Obviously, σ is surjective, and σ is also injective. In fact, if $sef = 0$, then it follows that $0 = sefg = se^2 = se$. So $se = 0$ and σ is an isomorphism. Hence also Sf is S-projective. Furthermore $d := gf = d^2 \in T$ is an idempotent. For this idempotent we have

$$Sf = Sfd \subseteq Hd = (Hg)f \subseteq Sf,$$

where we used that $Hg \subseteq S$ and $Sf \subseteq H$. Hence $Sf = Hd$ and Hd is an S-direct summand of H.

2) The proof is similar to that of 1). First we have the epimorphism

$$\tau : dT \ni dt \mapsto fdt = ft \in fT.$$

If now $fdt = 0$, then it follows that

$$0 = gfdt = d^2t = dt,$$

hence τ is also injective. Then together with dT, the T-module fT is also projective. Further we have

$$eH = fgH \subseteq fT = efT \subseteq eH,$$

hence $eH = fT \subseteq^{\oplus} H_T$. $\square$

Until now we considered H as an S-T-bimodule. We set

$$S' := \operatorname{End}(H_T), \quad T' := \operatorname{End}(_SH).$$

Then H is also a T'-left and a S'-right module. We ask how S and S' are related. If $s \in S$, then the mapping

$$\widehat{s} : H \ni h \mapsto sh \in H$$

is in S'. In particular, $\widehat{e} \in S'$ and $\widehat{e}$ is again an idempotent. Since $ef = f \neq 0$, there exists $a \in A$ with $ef(a) = f(a) \neq 0$, so also $\widehat{e} \neq 0$. As an element of S' we denote $\widehat{e}$ again by e. Similarly, we can assume that $d \in T'$.

Corollary 3.2. *If* $0 \neq f \in \operatorname{Hom}_R(A, M)$ *is regular, then*

1) $S'f \neq 0$ *is an S'-projective direct summand of* $'_S \operatorname{Hom}_R(A, M)$.

2) $fT' \neq 0$ *is an T'-projective direct summand of* $\operatorname{Hom}_R(A, M)'_T$.

Proof. The same as the proof of Theorem 3.1. $\square$

Remark. Corollary 3.2 can also be proved using later results. If f is regular in $H := \operatorname{Hom}_R(A, M)$, then (Lemma 2.1) f is a regular element in the modules $_SH$ and H_T and the result follows from the module case (Theorem 3.1).

We will return later to the rings S' and T'. They are important for the existence of certain substructures that we will consider later.

4 Existence and Properties of Reg(A, M)

We will show that there exists a largest regular S-T-submodule of $H = \mathrm{Hom}_R(A, M)$, denoted by $\mathrm{Reg}(A, M)$. Here "largest" means that any other regular S-T-submodule of H is contained in $\mathrm{Reg}(A, M)$.

For the proof we need the following fact.

Lemma 4.1. *Let* $f \in \mathrm{Hom}_R(A, M)$ *and* $g \in \mathrm{Hom}_R(M, A)$. *If* $f - fgf$ *is regular, then* f *is regular.*

Proof. Since $f - fgf$ is regular, there exists $h \in \mathrm{Hom}_R(M, A)$ such that

$$(f - fgf)h(f - fgf) = f - fgf. \tag{9}$$

Let
$$k := h - gfh - hfg + gfhfg + g \in \mathrm{Hom}_R(M, A)$$

with g from (9). An easy computation shows that $fkf = f$. In fact,

$$fkf = fhf - fgfhf - fhfgf + fgfhfgf + fgf = (f - fgf)h(f - fgf) + fgf$$

and by (9)
$$fkf = f - fgf + fgf = f. \qquad \square$$

Recall that a subset X of $\mathrm{Hom}_R(A, M)$ is regular if every element of X is regular.

Definition 4.2. Let

$$\mathrm{Reg}(A, M) := \{f \in \mathrm{Hom}_R(A, M) \mid SfT \text{ is regular}\},$$

where SfT is the S-T-submodule of $\mathrm{Hom}_R(A, M)$ generated by f.

Theorem 4.3. $\mathrm{Reg}(A, M)$ *is the largest regular* S-T-submodule of $\mathrm{Hom}_R(A, M)$.

Proof. The proof is achieved in four steps.

(a) Let $f \in H$ and assume that SfT is regular. If $\varphi \in SfT$, then $S\varphi T \subseteq SfT$, hence also $S\varphi T$ is regular and $\varphi \in \mathrm{Reg}(A, M)$. Since φ was an arbitrary element of SfT, it follows that $SfT \subseteq \mathrm{Reg}(A, M)$.

(b) Since $\mathrm{Reg}(A, M)$ is the sum of modules of the form SfT, it is closed under left multiplication by S and right multiplication by T.

(c) We show now that $\mathrm{Reg}(A, M)$ is closed under addition. Let $f_1, f_2 \in \mathrm{Reg}(A, M)$. We have to show that $S(f_1 + f_2)T$ is regular. An element of $S(f_1 + f_2)T$ is of the form

$$\sum_{i=1}^n s_i(f_1 + f_2)t_i = \sum_{i=1}^n s_i f_1 t_i + \sum_{i=1}^n s_i f_2 t_i.$$

Set $\varphi_1 = \sum_{i=1}^{n} s_i f_1 t_i$ and $\varphi_2 = \sum_{i=1}^{n} s_i f_2 t_i$. Then $\varphi_1 \in S f_1 T$ is regular and there exists $g \in \operatorname{Hom}_R(M, A)$ such that $\varphi_1 g \varphi_1 = \varphi_1$. Now consider

$$(\varphi_1 + \varphi_2) - (\varphi_1 + \varphi_2) g (\varphi_1 + \varphi_2) = \varphi_2 - \varphi_1 g \varphi_2 - \varphi_2 g \varphi_1 - \varphi_2 g \varphi_2, \qquad (10)$$

where we used that $\varphi_1 g \varphi_1 - \varphi_1 = 0$. Since $\varphi_1 g \in S$, $g \varphi_1, \varphi_2 g \in T$, the element in (10) is in $S \varphi_2 T$. We now use that $S \varphi_2 T$ is regular which is true by (a). We can then apply Lemma 4.1 to the element in (10) with $\varphi_1 + \varphi_2$ in place of f, and obtain that $\varphi_1 + \varphi_2$ is regular. Since $\varphi_1 + \varphi_2$ was an arbitrary element of $S(f_1 + f_2)T$, this submodule is regular. So $S(f_1 + f_2)T \subseteq \operatorname{Reg}(A, M)$ and in particular, $f_1 + f_2 \in \operatorname{Reg}(A, M)$.

(d) Finally we show that $\operatorname{Reg}(A, M)$ is the largest regular S-T-submodule of H. Assume that Λ is a regular S-T-submodule of H and $f \in \Lambda$. Then SfT is a regular S-T-submodule of H and by definition of $\operatorname{Reg}(A, M)$ we have $SfT \subseteq \operatorname{Reg}(A, M)$, in particular $f \in \operatorname{Reg}(A, M)$. This means that $\Lambda \subseteq \operatorname{Reg}(A, M)$ and $\operatorname{Reg}(A, M)$ is the largest regular S-T-submodule of H. $\qquad \square$

From Corollary 1.4 we can get examples for the extreme case $\operatorname{Hom}_R(A, M) = \operatorname{Reg}(A, M)$ but the conditions in Corollary 1.4 are by no means necessary. For example $\operatorname{Hom}_{\mathbb{Z}}(\mathbb{Q}, \mathbb{Q}) = \operatorname{Reg}(\mathbb{Q}, \mathbb{Q}) \cong \mathbb{Q}$ and $\mathbb{Q}_{\mathbb{Z}}$ is not semisimple. An example of the other extreme is $\operatorname{Reg}(\mathbb{Z}, \mathbb{Z}) = 0$ in spite of the fact that $\operatorname{Hom}_{\mathbb{Z}}(\mathbb{Z}, \mathbb{Z}) \cong \mathbb{Z}$ contains the regular elements ± 1.

We state an immediate consequence of Corollary 1.5.

Corollary 4.4. *Let A and M be modules such that no nonzero summand of A is isomorphic with a summand of M, then $\operatorname{Reg}(A, M) = 0$.*

We now come to interesting properties of the bimodule $\operatorname{Reg}(A, M)$. In order to avoid repetitions we abstract the essential step in the proofs of several theorems in a lemma.

Lemma 4.5. *Let H be a left S-module and N a submodule with the property that every cyclic submodule of N is a direct summand of H. Then the following statements hold.*

1) *Every finitely generated S-submodule U of N is a direct summand of H and a direct sum of cyclic submodules.*

2) *Every countably generated submodule of N is the direct sum of cyclic submodules.*

Proof. 1) The proof is by induction on the number n of generators $f_1, \ldots, f_n \in N$ of the submodule $U := \sum_{i=1}^{n} S f_i$ of N. The start $n = 1$ of the induction is hypothesis. Let $n > 1$ and set $V := \sum_{i=1}^{n-1} S f_i$. Then $U = V + S f_n$ and both V and U are submodules of N. By induction hypothesis V is a direct sum of cyclic submodules and a direct summand of H. Hence there is an idempotent

$d \in \mathrm{End}(_S H)$ such that $Hd = V$. We further have that $H = Hd \oplus H(1 - d)$ and $f_n = f_n d + f_n(1 - d)$. Since $f_n d \in Hd = V$, it follows that

$$U = V + Sf_n = V + S(f_n d + f_n(1 - d)) = V \oplus Sf_n(1 - d).$$

This shows that U is a direct sum of cyclic submodules. It remains to show that U is a direct summand of H. Now $Sf_n(1 - d) \subseteq Sf_n + Sf_n d \subseteq N + V = N$ and by hypothesis $Sf(1 - d)$ is again a direct summand of H. Hence there exists an idempotent $e \in \mathrm{End}(_S H)$ such that $Sf_n(1 - d) = He$. Thus $U = V \oplus Sf_n(1 - d)$ is the direct sum of the two modules V and $He = Sf_n(1 - d) \subseteq H(1 - d)$. Intersecting $H = He \oplus H(1 - e)$ with $H(1 - d)$ we obtain $H(1 - d) = He \oplus (H(1 - d) \cap H(1 - e))$. Consequently,

$$
\begin{aligned}
H &= Hd \oplus H(1 - d) = Hd \oplus He \oplus (H(1 - d) \cap H(1 - e)) \\
&= U \oplus (H(1 - d) \cap H(1 - e)).
\end{aligned}
$$

So U is a direct summand of H, and the proof is complete.

2) Let $U = \sum_{n=1}^{\infty} Sf_n \subseteq N$. By the proof of 1), for $n = 1, 2, \ldots$, we have $Sf_1 + \cdots + Sf_n = (Sf_1 + \cdots + Sf_{n-1}) \oplus C_n$ for some submodule C_n. Clearly $\sum_{n=1}^{\infty} C_n = \bigoplus_{n=1}^{\infty} C_n = U$. $\qquad\square$

Theorem 4.6. *Every finitely or countably generated S-submodule of $\mathrm{Reg}(A, M)$ is a direct sum of cyclic S-projective submodules that are isomorphic to left ideals of S that are direct summands of S.*

Every finitely generated S-submodule L of $\mathrm{Reg}(A, M)$ is S-projective and a direct summand of $_S \mathrm{Hom}_R(A, M)$.

The analogous results hold for $\mathrm{Reg}(A, M)$ as a right T-module.

Proof. By Theorem 3.1 every cyclic S-submodule of $\mathrm{Reg}(A, M)$ is a projective direct summand of $H = \mathrm{Hom}_R(A, M)$. Hence Lemma 4.5 applies and it is clear that the submodules in question are projective in this case. $\qquad\square$

We will make use of a lemma for a general module $_S H$.

Lemma 4.7. *Let $_S H$ be a left S-module and suppose that $_S H$ contains no infinite direct sum of submodules and every cyclic submodule of H is a direct summand. Then H is finitely generated and semisimple.*

Proof. By Lemma 4.5 we have that every finitely generated submodule is a direct summand of H.

We show next that every submodule of H is finitely generated. By way of contradiction assume that N is a submodule that is not finitely generated. Then N contains elements $n_1, n_2, \ldots$ such that

$$Sn_1 \subsetneqq Sn_1 + Sn_2 \subsetneqq \cdots \subsetneqq \sum_{i=1}^{k} Sn_i \subsetneqq \sum_{i=1}^{k+1} Sn_i \subsetneqq \cdots$$

As every finitely generated submodule of H is a direct summand of H and hence of every submodule that contains it, we obtain decompositions

$$Sn_1 + Sn_2 = Sn_1 \oplus K_1,$$
$$Sn_1 + Sn_2 + Sn_3 = (Sn_1 + Sn_2) \oplus K_2,$$
$$\vdots$$
$$Sn_1 + \cdots + Sn_{k+1} = (Sn_1 + \cdots + Sn_k) \oplus K_k$$
$$\vdots$$

So $\sum_{i=1}^{\infty} Sn_i = n_1 R \oplus K_1 \oplus K_2 \oplus \cdots$ which is an infinite direct sum contradicting the hypothesis. So every submodule is finitely generated and therefore a direct summand of H. In particular, $H = \mathrm{Soc}(H) \oplus N$ for some submodule N where $\mathrm{Soc}(H)$ is the sum of all simple submodules of H. Suppose that Sm is a nonzero cyclic submodule of N. There is a maximal left ideal I of S containing $\mathrm{Ann}_S(m)$ and $Im \subseteq Sm$ is a cyclic submodule of H. Hence $Sm = Im \oplus K$ for some K and $K \cong Sm/Im \cong S/I$ is simple. But then $K \subseteq \mathrm{Soc}(H) \cap N = 0$, a contradiction. $\quad\square$

Corollary 4.8. *Suppose that* $\mathrm{Reg}(A, M)$ *contains no infinite direct sums of S-submodules. Then* $\mathrm{Reg}(A, M)$ *is the direct sum of finitely many simple projective S-modules,* $\mathrm{Hom}_R(A, M) = \mathrm{Reg}(A, M) \oplus U$ *for some S-submodule U and U contains no nonzero regular S-T-submodule. Every cyclic S-submodule of* $\mathrm{Reg}(A, M)$ *is isomorphic to a left ideal of S that is a direct summand of S.*

There is an analogous result for $\mathrm{Hom}_R(A, M)_T$.

Proof. By Theorem 4.6 and Lemma 4.7 the S-module $\mathrm{Reg}(A, M)$ must be finitely generated and a finite direct sum of simple submodules. By Lemma 4.5 the submodule $\mathrm{Reg}(A, M)$ is a direct summand of H. Finally, by definition of $\mathrm{Reg}(A, M)$, the S-submodule U contains no nonzero regular S-T-submodule. $\quad\square$

Corollary 4.9. *Suppose that* $\mathrm{Hom}_R(A, M)$ *is regular. Then every finitely generated S-submodule L of* $\mathrm{Hom}_R(A, M)$ *is S-projective and a direct summand of* $_S\mathrm{Hom}_R(A, M)$. *Furthermore, L is the direct sum of finitely many cyclic S-projective submodules that are isomorphic to left ideals of S.*

Under certain finiteness conditions we can obtain much more specific information on $\mathrm{Reg}(A, M)$.

Theorem 4.10. *Suppose that* $\mathrm{Hom}_R(A, M)$ *is regular and contains no infinite direct sums of S-submodules. Then* $\mathrm{Hom}_R(A, M)$ *is a finitely generated semisimple S-submodule. The simple direct summands of* $\mathrm{Hom}_R(A, M)$ *are S-projective and isomorphic to left ideals of S.*

Proof. It follows from the hypothesis that $H := \mathrm{Hom}_R(A, M)$ is regular that every cyclic submodule is a direct summand. The fact that H contains no infinite direct sums implies that every S-submodule of H is finitely generated. Hence every submodule of $_SH$ is a direct summand of H. It follows that $\mathrm{Rad}(H) = 0$ and this together with the fact that every submodule of H is a direct summand implies that H is a direct sum of simple modules, i.e., H is semisimple. (See Chapter I). □

The following result generalizes part of [37, Theorem 1.11]. A ring R is **left (right) perfect** if every left (right) R-module has a projective cover.

Theorem 4.11. *Suppose that* $S := \mathrm{End}(M_R)$ *is left perfect. Then* $\mathrm{Reg}(A, M)$ *is S-projective. If* $T := \mathrm{End}(A_R)$ *is right perfect, then* $\mathrm{Reg}(A, M)$ *is T-projective.*

Proof. As S is left perfect, by [9, page 68, P(3')] direct limits of S-projective modules are S-projective. The S-module $\mathrm{Reg}(A, M)$ is the direct limit of its finitely generated submodules and by Theorem 4.6 these are all S-projective. Hence their direct limit $\mathrm{Reg}(A, M)$ is S-projective. □

5 The Transfer Rule

There is a connection between regular elements in $\mathrm{Hom}_R(A, M)$ and regular elements in $\mathrm{Hom}_R(M, A)$ that is as surprising as it is simple.

If $f \in \mathrm{Hom}_R(A, M)$ and $g \in \mathrm{Hom}_R(M, A)$ with $fgf = f$, then we call (f, g) a **regular pair**. Similarly, if $h \in \mathrm{Hom}_R(M, A)$ and $k \in \mathrm{Hom}_R(A, M)$ with $hkh = h$, then (h, k) is a regular pair. If (f, g) is a regular pair, then also (gfg, f) is a regular pair because

$$gfg \cdot f \cdot gfg = gfg$$

but now with the regular homomorphism $gfg \in \mathrm{Hom}_R(M, A)$.

We will show that with these formal rules we can obtain all regular elements in $\mathrm{Hom}_R(M, A)$ from those in $\mathrm{Hom}_R(A, M)$.

Definition 5.1. If (f, g) is a regular with pair $f \in \mathrm{Hom}_R(A, M)$, then we call

$$\mathrm{trf} : (f, g) \mapsto (gfg, f)$$

the **transfer rule** or the **transfer** and we also write $\mathrm{trf}(f, g) = (gfg, f)$.

Applying trf twice to the pair (f, g) we obtain

$$(f, g) \overset{\mathrm{trf}}{\mapsto} (gfg, f) \overset{\mathrm{trf}}{\mapsto} (fgfgf, gfg) = (f, gfg).$$

Hence applying trf twice to (f, g) brings us to (f, gfg). Thus we come back to f with which we started, but now paired with the quasi-inverse gfg in place of g.

Note that transfer produced a regular pair (f, gfg) whose entries are mutual quasi-inverses.

If $f \in H$ is regular and g_1, g_2 are quasi-inverses of f, then

$$\mathrm{trf}(f, g_1) = (g_1 f g_1, f) \text{ and } \mathrm{trf}(f, g_2) = (g_2 f g_2, f).$$

The following example shows that $g_1 f g_1 \neq g_2 f g_2$ is possible.

Example 5.2. Let $A = \mathbb{Z}a \oplus \mathbb{Z}b$ where $\mathrm{ord}(a) = \infty$, $\mathrm{ord}(b) = p$, and let $M = A$. Then also $M = \mathbb{Z}(a+b) \oplus \mathbb{Z}b$. Let $f : A \ni ra + sb \mapsto ra \in M$, $g_1 : M \ni ra + sb \mapsto ra \in A$, and $g_2 : M \ni ra + sb = r(a+b) + (s-r)b \mapsto r(a+b) \in A$. It is easily checked that $fg_1f = f$ and $fg_2f = f$ but $g_1fg_1 \neq g_2fg_2$, in fact $g_1fg_1(a) = a$ while $g_2fg_2(a) = a + b$.

Theorem 5.3. *If* trf *is applied to all regular pairs* (f,g) *with* $f \in \mathrm{Hom}_R(A,M)$, *then the set of first entries in* $\mathrm{trf}(f,g) = (gfg, f)$ *is the set of all regular elements in* $\mathrm{Hom}_R(M,A)$.

Proof. Let (h,k) be a regular pair with $h \in \mathrm{Hom}_R(M,A)$ and $k \in \mathrm{Hom}_R(A,M)$. Then $\mathrm{trf}(h,k) = (khk, h)$ is again a regular pair with $khk \in \mathrm{Hom}_R(A,M)$. Applying trf again, then we get (h, khk). Hence every regular element in $\mathrm{Hom}_R(M,A)$ is obtained by an application of trf to the element $khk \in \mathrm{Hom}_R(A,M)$. $\qquad\square$

Let (f,g) be a regular pair. Recall (Theorem 1.2) that we have decompositions

$$A = \mathrm{Im}(gf) \oplus \mathrm{Ker}(f), \quad M = \mathrm{Im}(f) \oplus \mathrm{Ker}(fg), \quad \text{and} \quad \mathrm{Im}(f) = \mathrm{Im}(fg).$$

Using the regular pair (gfg, f) obtained by transfer, one gets further relationships. To wit:

1) $\mathrm{Im}(gf) = \mathrm{Im}(gfg)$, $\mathrm{Ker}(gfg) = \mathrm{Ker}(fg)$,

2) $\mathrm{Im}(g) = \mathrm{Im}(gf) \oplus (\mathrm{Im}(g) \cap \mathrm{Ker}(f)) \subseteq A$.

Proof. We have

$$A = \mathrm{Ker}(f) \oplus \mathrm{Im}(gf) = \mathrm{Ker}(f) \oplus \mathrm{Im}(gfg),$$

with $\mathrm{Im}(gfg) \subseteq \mathrm{Im}(gf)$. It follows that $\mathrm{Im}(gf) = \mathrm{Im}(gfg)$.
We have

$$M = \mathrm{Im}(f) \oplus \mathrm{Ker}(fg) = \mathrm{Im}(f) \oplus \mathrm{Ker}(gfg),$$

with $\mathrm{Ker}(gf) \subseteq \mathrm{Ker}(gfg)$. It follows that $\mathrm{Ker}(gf) = \mathrm{Ker}(gfg)$.
We have $\mathrm{Im}(gf) \subseteq \mathrm{Im}(g)$ and $A = \mathrm{Ker}(f) \oplus \mathrm{Im}(gf)$, hence $\mathrm{Im}(g) = \mathrm{Im}(gf) \oplus (\mathrm{Im}(g) \cap \mathrm{Ker}(f))$. $\qquad\square$

Question 5.4. How does transfer act on Reg?

6   Inherited Regularity

We are interested in regularity in $\mathrm{Hom}_R(A,M)$ for R-modules A, M and ask how regularity interacts with related homomorphism groups. Suppose that A' and M' are two more R-modules and that homomorphisms

$$\phi : A \to A', \quad \mu : M \to M'$$

are given. We then have induced maps on homomorphism groups as follows.

$$\phi^* : \mathrm{Hom}_R(A', M) \ni f \ \mapsto \ f \circ \phi \in \mathrm{Hom}_R(A, M),$$
$$\mu_* : \mathrm{Hom}_R(A, M) \ni f \ \mapsto \ \mu \circ f \in \mathrm{Hom}_R(A, M'),$$
$$\phi_* : \mathrm{Hom}_R(M, A) \ni f \ \mapsto \ \phi \circ f \in \mathrm{Hom}_R(M, A'),$$
$$\mu^* : \mathrm{Hom}_R(M', A) \ni f \ \mapsto \ f \circ \mu \in \mathrm{Hom}_R(M, A).$$

Proposition 6.1. *Let $f \in \mathrm{Hom}_R(A', M)$ and assume the existence of the maps listed above. Then the following statements hold.*

1) *If ϕ is surjective and $\phi^*(f)$ is regular, then f is regular.*

2) *If μ is injective and $\mu_*(f)$ is regular, then f is regular.*

3) *If ϕ_* is surjective and f is regular, then $\phi^*(f)$ is regular.*

4) *If μ^* is surjective and f is regular, then $\mu_*(f)$ is regular.*

Proof. 1) Suppose that ϕ is surjective and $\phi^*(f)$ is regular. Then there exists $g \in \mathrm{Hom}_R(M, A)$ such that $\phi^*(f)g\phi^*(f) = \phi^*(f)$. Equivalently, $f\phi gf\phi = f\phi$. Since ϕ is surjective, it can be canceled, and we obtain the regularity equation $f(\phi g)f = f$.

2) Suppose that μ is injective and $\mu_*(f)$ is regular. Then there is $g \in \mathrm{Hom}_R(M', A)$ such that $\mu_*(f)g\mu_*(f) = \mu_*(f)$. Equivalently, $\mu fg\mu f = \mu f$. Since μ is injective, it can be canceled and we obtain the regularity equation $f(g\mu)f = f$.

3) Suppose that ϕ_* is surjective and $fgf = f$. Since ϕ_* is surjective there exists $h \in \mathrm{Hom}_R(M, A)$ such that $g = \phi_*(h) = \phi h$. We conclude that $f\phi hf\phi = f\phi$, or, equivalently, $\phi^*(f)h\phi^*(f) = \phi^*(f)$.

4) Suppose that μ^* is surjective and $fgf = f$ for a $g \in \mathrm{Hom}_R(M, A')$. Since μ^* is surjective, there is $h \in \mathrm{Hom}_R(M', A)$ such that $g = \mu^*(h) = h\mu$. We conclude that $\mu fh\mu f = \mu f$, or, equivalently, $\mu_*(f)h\mu_*(f) = \mu_*(f)$. $\square$

The claims of the following corollary are immediate consequences of Proposition 6.1.1 and 6.1.2.

Corollary 6.2.

1) *If $\phi \in \mathrm{Hom}_R(A, A')$ is surjective and $\mathrm{Hom}_R(A, M)$ is regular, then $\mathrm{Hom}_R(A', M)$ is regular.*

2) *If $\mu \in \mathrm{Hom}_R(M, M')$ is injective and $\mathrm{Hom}_R(A, M')$ is regular, then $\mathrm{Hom}_R(A, M)$ is regular.*

An interesting special case of Corollary 6.2 is as follows.

Corollary 6.3.

1) *If $B \subseteq A$ and $\mathrm{Hom}_R(A, M)$ is regular, then also $\mathrm{Hom}_R(A/B, M)$ is regular.*

2) *If $N \subseteq M$, and $\mathrm{Hom}_R(A, M)$ is regular, then $\mathrm{Hom}_R(A, N)$ is regular.*

Proof. Apply Corollary 6.2 with $\phi : A \twoheadrightarrow A/B$ and $\mu : N \hookrightarrow M$. $\qquad\square$

We will now study the case of a map $\phi : A \to A'$ more closely and check what hypotheses are needed in order to get results on the relationships of $\text{Reg}(A, M)$ and $\text{Reg}(A', M)$. We have

$$\phi : A \to A' \quad \text{induces} \quad \phi^* : \text{Hom}_R(A', M) \to \text{Hom}_R(A, M). \qquad (11)$$

The abelian group $H := \text{Hom}_R(A, M)$ is an S-T-bimodule where $S := \text{End}(M_R)$ and $T := \text{End}(A_R)$ while $H' := \text{Hom}_R(A', M)$ is an S-T'-bimodule where $S := \text{End}(M_R)$ as before and $T' := \text{End}(A'_R)$. Clearly, the map ϕ^* is an S-module morphism. Suppose that

$$\forall t \in \text{End}(A_R), \exists t' \in \text{End}(A'_R) \text{ such that } t'\phi = \phi t \qquad (12)$$

or

$$\forall t' \in \text{End}(A'_R), \exists t \in \text{End}(A_R) \text{ such that } t'\phi = \phi t. \qquad (13)$$

Then in either case

$$\forall f \in \text{Hom}_R(A', M) : \phi^*(f)t = \phi^*(ft').$$

Proposition 6.4. *Let $\phi : A \to A'$ and assume that*

$$\forall g \in \text{Hom}_R(M, A'), \exists h \in \text{Hom}_R(M, A) \text{ such that } g = \phi h, \qquad (14)$$

and assume (12). Then

$$\phi^*(\text{Reg}(A', M)) \subseteq \text{Reg}(A, M).$$

Proof. Let $f \in \text{Reg}(A', M)$. Then f is regular and there is $g \in \text{Hom}_R(M, A')$ such that $fgf = f$. By (14) we get $g = \phi h$ and hence $\phi^*(f)h\phi*(f) = f\phi h f\phi = fgf\phi = f\phi = \phi^*(f)$, showing that $\phi^*(f) \in \text{Hom}_R(A, M)$ is regular. By definition of Reg the claim is established if we can show that $S\phi^*(\text{Reg}(A', M))T \subseteq \phi^*(\text{Reg}(A', M))$. Since ϕ^* is an S-homomorphism, we have $S\phi^*(\text{Reg}(A', M)) = \phi^*(S \text{Reg}(A', M)) \subseteq \phi^*(\text{Reg}(A', M))$. Now let $f \in \text{Reg}(A', M)$ and $t \in \text{End}(A_R)$. By (12) there is $t' \in \text{End}(A'_R)$ such that $\phi t' = t\phi$. Then $\phi^*(f) \cdot t = f\phi t = ft'\phi = \phi^*(ft')$ which is regular because $ft' \in \text{Reg}(A', M)$. $\qquad\square$

Proposition 6.5. *Let $\phi : A \to A'$ and assume that ϕ is surjective and that (13) holds. Then*

$$(\phi^*)^{-1}[\text{Reg}(A, M)] \subseteq \text{Reg}(A', M).$$

Proof. Let $f \in \text{Hom}_R(A', M)$ such that $\phi^*(f) \in \text{Reg}(A, M)$. Then $\phi^*(f)$ is regular and there is $g \in \text{Hom}_R(M, A)$ such that $\phi^*(f)g\phi^*(f) = \phi^*(f)$ or equivalently, $f\phi gf\phi = f\phi$. Since ϕ is surjective, we obtain that $f(\phi g)f = f$ showing that f is regular. By definition of Reg the claim is established if we can show that $S(\phi^*)^{-1}[\text{Reg}(A', M)]T \subseteq (\phi^*)^{-1}[\text{Reg}(A', M)]$. Since ϕ^* is an S-homomorphism, for $s \in S$ we have $\phi^*(sf) = s\phi^*(f) \in \text{Reg}(A, M))$, hence $sf \in (\phi^*)^{-1}[\text{Reg}(A', M)]$. Now let $t' \in \text{End}(A'_R)$. By (13) there is $t \in \text{End}(A_R)$ such that $\phi t = t'\phi$. Then $\phi^*(ft') = ft'\phi = f\phi t = \phi^*(f)t \in \text{Reg}(A, M)$. $\qquad\square$

We now study the case of a map $\phi : M \to M'$ more closely and check what hypotheses are needed in order to get results on the relationships of $\mathrm{Reg}(A, M)$ and $\mathrm{Reg}(A, M')$. We have

$$\mu : M \to M' \quad \text{induces} \quad \mu_* : \mathrm{Hom}_R(A, M) \to \mathrm{Hom}_R(A, M'). \qquad (15)$$

The $H := \mathrm{Hom}_R(A, M)$ is an S-T-bimodule where $S := \mathrm{End}(M_R)$ and $T := \mathrm{End}(A_R)$ while $H' := \mathrm{Hom}_R(A, M')$ is an S'-T-bimodule where $S' := \mathrm{End}(M'_R)$ and $T = \mathrm{End}(A_R)$. Clearly, the map μ^* is an T-module morphism. Suppose that

$$\forall s \in \mathrm{End}(M_R), \exists s' \in \mathrm{End}(M'_R) \text{ such that } s'\mu = \mu s \qquad (16)$$

or

$$\forall s' \in \mathrm{End}(M'_R), \exists s \in \mathrm{End}(M_R) \text{ such that } s'\mu = \mu s. \qquad (17)$$

Then in either case

$$\forall f \in \mathrm{Hom}_R(A, M) : s'\mu_*(f) = \mu_*(sf).$$

Proposition 6.6. *Let $\mu : M \to M'$ and assume that*

$$\forall g \in \mathrm{Hom}_R(M, A), \exists h \in \mathrm{Hom}_R(M', A) \text{ such that } g = h\mu, \qquad (18)$$

and that (16) holds. Then

$$\mu_*(\mathrm{Reg}(A, M)) \subseteq \mathrm{Reg}(A, M').$$

Proof. Let $f \in \mathrm{Reg}(A, M)$. Then f is regular and there is $g \in \mathrm{Hom}_R(M, A)$ such that $fgf = f$. By (18) we get $\mu_*(fgf) = \mu fgf = \mu fh\mu f = \mu f$, showing that $\mu_*(f) = \mu f \in \mathrm{Hom}_R(A, M')$ is regular. By definition of Reg the claim is established if we can show that $S'\mu_*(\mathrm{Reg}(A, M')) \subseteq \mu_*(\mathrm{Reg}(A, M'))$. Since μ_* is a T-homomorphism, we have $\mu_*(\mathrm{Reg}(A, M))T = \mu_*(\mathrm{Reg}(A, M)T) \subseteq \mu_*(\mathrm{Reg}(A, M))$. Now let $f \in \mathrm{Reg}(A, M)$ and $s' \in S' = \mathrm{End}(M'_R)$. By (17) there is $s \in \mathrm{End}(M_R)$ such that $s'\mu = \mu s$. Then $s' \cdot \mu_*(f) = s'\mu f = \mu sf = \mu_*(sf)$ which is regular because $sf \in \mathrm{Reg}(A, M)$. $\qquad \square$

Proposition 6.7. *Let $\mu : M \to M'$ and assume that μ is injective and that (16) holds. Then*

$$(\mu_*)^{-1}[\mathrm{Reg}(A, M')] \subseteq \mathrm{Reg}(A, M).$$

Proof. Let $f \in \mathrm{Hom}_R(A, M)$ such that $\mu_*(f) \in \mathrm{Reg}(A, M')$. Then $\mu_*(f)$ is regular and there is $g \in \mathrm{Hom}_R(M', A)$ such that $\mu_*(f)g\mu_*(f) = \mu_*(f)$ or equivalently, $\mu fg\mu f = \mu f$. Since μ is injective, we obtain that $f(g\mu)f = f$ showing that f is regular. By definition of Reg the claim is established if we can show that $S'(\mu_*)^{-1}[\mathrm{Reg}(A, M')]T \subseteq (\mu_*)^{-1}[\mathrm{Reg}(A, M')]$. Since μ_* is a T-homomorphism, for $t \in T$ we have $\mu_*(ft) = \mu_*(f)t \in \mathrm{Reg}(A, M'))$, hence $ft \in (\mu_*)^{-1}[\mathrm{Reg}(A, M')]$. Now let $s' \in \mathrm{End}(M'_R)$. By (17) there is $s \in \mathrm{End}(M_R)$ such that $s'\mu = \mu s$. Then $\mu_*(sf) = \mu sf = s'\mu f = s'\mu_*(f) \in \mathrm{Reg}(A, M')$. $\qquad \square$

We now apply our results to the short exact sequence

$$B \overset{\iota}{\rightarrowtail} A \overset{\phi}{\twoheadrightarrow} A/B$$

where ι is the insertion and ϕ is the natural epimorphism. We obtain the exact sequence

$$\operatorname{Hom}_R(A/B, M) \overset{\phi^*}{\rightarrowtail} \operatorname{Hom}_R(A, M) \overset{\iota^*}{\twoheadrightarrow} \operatorname{Hom}_R(B, M).$$

Hence we have

- $\phi^*(\operatorname{Reg}(A/B, M)) \subseteq \operatorname{Hom}_R(A, M)$,
- $\iota^*(\operatorname{Reg}(A, M)) \subseteq \operatorname{Hom}_R(B, M)$,
- $(\phi^*)^{-1}[\operatorname{Reg}(A, M)] \subseteq \operatorname{Hom}_R(A/B, M)$,
- $(\iota^*)^{-1}[\operatorname{Reg}(B, M)] \subseteq \operatorname{Hom}_R(A, M)$.

Theorem 6.8. *Let $B \overset{\iota}{\rightarrowtail} A \overset{\phi}{\twoheadrightarrow} A/B$ be given.*

1) *Suppose that $\forall t' \in \operatorname{End}((A/B)_R), \exists t \in \operatorname{End}(A_R) : t'\phi = \phi t$. Then*

$$(\phi^*)^{-1}[\operatorname{Reg}(A, M)] \subseteq \operatorname{Reg}(A/B, M).$$

2) *Suppose that $\iota_* : \operatorname{Hom}_R(M, B) \to \operatorname{Hom}_R(M, A)$ is surjective and $\forall t \in \operatorname{End}(B_R), \exists t' \in \operatorname{End}(A_R) : t'\iota = \iota t$. Then*

$$(\iota^*)^{-1}[\operatorname{Reg}(B, M)] \subseteq \operatorname{Reg}(A, M).$$

Theorem 6.9. *Let $N \overset{\iota}{\rightarrowtail} M \overset{\mu}{\twoheadrightarrow} M/N$ be given.*

1) *Suppose that $\mu^* : \operatorname{Hom}_R(M/N, A) \to \operatorname{Hom}(M, A)$ is surjective and N is fully invariant in M. Then*

$$\mu_*(\operatorname{Reg}(A, M)) \subseteq \operatorname{Reg}(A, M/N).$$

2) *Suppose that $\iota^* : \operatorname{Hom}_R(M, A) \to \operatorname{Hom}_R(N, A)$ is surjective and $\forall s \in \operatorname{End}(N_R), \exists s' \in \operatorname{End}(M_R) : s'\iota = \iota s$. Then*

$$\iota_*(\operatorname{Reg}(A, N)) \subseteq \operatorname{Reg}(A, M).$$

We next consider the crucial case when A and M are direct sums.
Let A and M be right R-modules and suppose that

$$A = A_1 \oplus A_2, \quad M = M_1 \oplus M_2. \tag{19}$$

It is a standard fact that

$$\text{Hom}_R(A, M)$$
$$\cong \text{Hom}_R(A_1, M_1) \oplus \text{Hom}_R(A_1, M_2) \oplus \text{Hom}_R(A_2, M_1) \oplus \text{Hom}_R(A_2, M_2).$$

We will study the relationship between regular elements of $\text{Hom}_R(A, M)$ and those of its additive subgroups $\text{Hom}_R(A_i, M_j)$. It is convenient to identify $\text{Hom}_R(A, M)$ with a group of matrices as follows.

Lemma 6.10. *Suppose that* $A = A_1 \oplus A_2$ *and* $M = M_1 \oplus M_2$. *Then* $\text{Hom}_R(A, M)$ *may be identified with the additive group of all matrices*

$$\begin{bmatrix} \xi_{11} & \xi_{21} \\ \xi_{12} & \xi_{22} \end{bmatrix}, \quad \begin{matrix} \xi_{11} \in \text{Hom}_R(A_1, M_1), & \xi_{21} \in \text{Hom}_R(A_2, M_1), \\ \xi_{12} \in \text{Hom}_R(A_1, M_2), & \xi_{22} \in \text{Hom}_R(A_2, M_2). \end{matrix}$$

A matrix acts on an element $a_1 + a_2$, *where* $a_i \in A_i$, *according to ordinary matrix multiplication,*

$$\begin{bmatrix} \xi_{11} & \xi_{21} \\ \xi_{12} & \xi_{22} \end{bmatrix} \begin{bmatrix} a_1 \\ a_2 \end{bmatrix} = \begin{bmatrix} \xi_{11}(a_1) + \xi_{21}(a_2) \\ \xi_{12}(a_1) + \xi_{22}(a_2) \end{bmatrix}.$$

The mapping

$$\text{Hom}_R(A, M) \ni \begin{bmatrix} \xi_{11} & \xi_{21} \\ \xi_{12} & \xi_{22} \end{bmatrix} \mapsto \xi_{ij} \in \text{Hom}_R(A_i, M_j)$$

is a surjective homomorphism of additive groups. Composition of functions is matrix multiplication.

Proof. Well-known and straightforward to check. □

We first relate regular elements in $\text{Hom}_R(A_i, M_j)$ with special regular matrices in $\text{Hom}_R(A, M)$.

Theorem 6.11 (Up and Down Theorem). *Suppose that* $A = A_1 \oplus A_2$ *and* $M = M_1 \oplus M_2$. *Then the following statements are true.*

1) $\xi_{11} \in \text{Hom}_R(A_1, M_1)$ *is regular if and only if* $\begin{bmatrix} \xi_{11} & 0 \\ 0 & 0 \end{bmatrix} \in \text{Hom}_R(A, M)$ *is regular.*

2) $\xi_{21} \in \text{Hom}_R(A_2, M_1)$ *is regular if and only if* $\begin{bmatrix} 0 & \xi_{21} \\ 0 & 0 \end{bmatrix} \in \text{Hom}_R(A, M)$ *is regular.*

3) $\xi_{12} \in \text{Hom}_R(A_1, M_2)$ *is regular if and only if* $\begin{bmatrix} 0 & 0 \\ \xi_{12} & 0 \end{bmatrix} \in \text{Hom}_R(A, M)$ *is regular.*

4) $\xi_{22} \in \mathrm{Hom}_R(A_2, M_2)$ *is regular if and only if* $\begin{bmatrix} 0 & 0 \\ 0 & \xi_{22} \end{bmatrix} \in \mathrm{Hom}_R(A, M)$ *is*
regular.

Proof. 1) Observe that for $\eta_{11} \in \mathrm{Hom}_R(M_1, A_1)$,

$$\xi_{11}\eta_{11}\xi_{11} = \xi_{11} \Rightarrow \begin{bmatrix} \xi_{11} & 0 \\ 0 & 0 \end{bmatrix} \begin{bmatrix} \eta_{11} & 0 \\ 0 & 0 \end{bmatrix} \begin{bmatrix} \xi_{11} & 0 \\ 0 & 0 \end{bmatrix} = \begin{bmatrix} \xi_{11} & 0 \\ 0 & 0 \end{bmatrix}.$$

Conversely, if

$$\begin{bmatrix} \xi_{11} & 0 \\ 0 & 0 \end{bmatrix} \begin{bmatrix} \eta_{11} & \eta_{21} \\ \eta_{12} & \eta_{22} \end{bmatrix} \begin{bmatrix} \xi_{11} & 0 \\ 0 & 0 \end{bmatrix} = \begin{bmatrix} \xi_{11} & 0 \\ 0 & 0 \end{bmatrix},$$

then by (23) also $\xi_{11}\eta_{11}\xi_{11} = \xi_{11}$.

2) By (32)

$$\begin{bmatrix} 0 & \xi_{21} \\ 0 & 0 \end{bmatrix} \begin{bmatrix} \eta_{11} & \eta_{21} \\ \eta_{12} & \eta_{22} \end{bmatrix} \begin{bmatrix} 0 & \xi_{21} \\ 0 & 0 \end{bmatrix} = \begin{bmatrix} 0 & \xi_{21}\eta_{12}\xi_{21} \\ 0 & 0 \end{bmatrix}. \tag{20}$$

Hence if $\xi_{21}\eta_{12}\xi_{21} = \xi_{21}$ for some $\eta_{21} \in \mathrm{Hom}_R(M_2, A_1)$, then choosing $\eta_{11} = \eta_{21} = \eta_{22} = 0$ in (20), we see that $\begin{bmatrix} 0 & \xi_{21} \\ 0 & 0 \end{bmatrix}$ is regular. Conversely, if $\begin{bmatrix} 0 & \xi_{21} \\ 0 & 0 \end{bmatrix}$
is regular, then the matrix in (20) equals $\begin{bmatrix} 0 & \xi_{21} \\ 0 & 0 \end{bmatrix}$ and it follows that $\xi_{21}\eta_{12}\xi_{21} = \xi_{21}$.

3) By (29)

$$\begin{bmatrix} 0 & 0 \\ \xi_{12} & 0 \end{bmatrix} \begin{bmatrix} \eta_{11} & \eta_{21} \\ \eta_{12} & \eta_{22} \end{bmatrix} \begin{bmatrix} 0 & 0 \\ \xi_{12} & 0 \end{bmatrix} = \begin{bmatrix} 0 & 0 \\ \xi_{12}\eta_{21}\xi_{12} & 0 \end{bmatrix}. \tag{21}$$

Hence if $\xi_{12}\eta_{21}\xi_{12} = \xi_{12}$ for some $\eta_{12} \in \mathrm{Hom}_R(M_1, A_2)$, then choosing $\eta_{11} = \eta_{12} = \eta_{22} = 0$ in (21), we see that $\begin{bmatrix} 0 & 0 \\ \xi_{12} & 0 \end{bmatrix}$ is regular. Conversely, if $\begin{bmatrix} 0 & 0 \\ \xi_{21} & 0 \end{bmatrix}$
is regular, then the matrix in (21) equals $\begin{bmatrix} 0 & 0 \\ \xi_{12} & 0 \end{bmatrix}$ and it follows that $\xi_{12}\eta_{21}\xi_{12} = \xi_{12}$.

4) By (38)

$$\begin{bmatrix} 0 & 0 \\ 0 & \xi_{22} \end{bmatrix} \begin{bmatrix} \eta_{11} & \eta_{21} \\ \eta_{12} & \eta_{22} \end{bmatrix} \begin{bmatrix} 0 & 0 \\ 0 & \xi_{22} \end{bmatrix} = \begin{bmatrix} 0 & 0 \\ 0 & \xi_{22}\eta_{22}\xi_{22} \end{bmatrix}. \tag{22}$$

Hence if $\xi_{22}\eta_{22}\xi_{22} = \xi_{22}$ for some $\eta_{22} \in \mathrm{Hom}_R(M_2, A_2)$, then choosing $\eta_{11} = \eta_{21} = \eta_{12} = 0$ in (22), we see that $\begin{bmatrix} 0 & 0 \\ 0 & \xi_{22} \end{bmatrix}$ is regular. Conversely, if $\begin{bmatrix} 0 & 0 \\ 0 & \xi_{22} \end{bmatrix}$
is regular, then the matrix in (22) equals $\begin{bmatrix} 0 & 0 \\ 0 & \xi_{22} \end{bmatrix}$ and it follows that $\xi_{22}\eta_{22}\xi_{22} = \xi_{22}$. $\qquad\square$

Corollary 6.12. *Suppose that $A = A_1 \oplus A_2$ and $M = M_1 \oplus M_2$. If $\mathrm{Reg}(A, M) = \mathrm{Hom}_R(A, M)$, then also $\mathrm{Reg}(A_i, M_j) = \mathrm{Hom}_R(A_i, M_j)$ for all $i, j \in \{1, 2\}$.*

We will see later that the converse of Corollary 6.12 is also true (Corollary 6.18)

A simple example shows that the restriction of a regular map to a direct summand need not be regular.

Example 6.13. Let $A = \mathbb{Z}u \oplus \mathbb{Z}v$ and $M = \mathbb{Z}u' \oplus \mathbb{Z}v'$ be free groups. Then also $M = \mathbb{Z}(u' + 2v') \oplus \mathbb{Z}v'$. Let $f \in \mathrm{Hom}(A, M)$ be the map given by $f(u) = 2(u' + 2v')$ and $f(v) = u' + 2v'$. Then $\mathrm{Im}(f) = \mathbb{Z}(u' + 2v')$ is a free direct summand of M and hence $\mathrm{Ker}(f)$ is also a summand of A, so f is regular. On the other hand $f(\mathbb{Z}u) = 2\,\mathrm{Im}(f)$ is not a direct summand of M and therefore $f \upharpoonright_{\mathbb{Z}u}$ is not regular.

We now study regularity in $\mathrm{Hom}_R(A_1 \oplus A_2, M_1 \oplus M_2)$ more generally. The following theorem generalizes a result of Goodearl ([14, Theorem 1.7]) and our proof owes much to the proof of Goodearl's Lemma 1.6.

Theorem 6.14. *Let $A = A_1 \oplus A_2$ and $M = M_1 \oplus M_2$ as before. We identify $\mathrm{Hom}_R(A, M)$ with a matrix group as in Lemma 6.10. Furthermore, we use the identifications*

$$S := \mathrm{End}_R(M) = \left\{ \begin{bmatrix} \mu_{11} & \mu_{21} \\ \mu_{12} & \mu_{22} \end{bmatrix} \mid \mu_{ij} \in \mathrm{Hom}_R(M_i, M_j) \right\}$$

and

$$T := \mathrm{End}_R(A) = \left\{ \begin{bmatrix} \alpha_{11} & \alpha_{21} \\ \alpha_{12} & \alpha_{22} \end{bmatrix} \mid \alpha_{ij} \in \mathrm{Hom}_R(A_i, A_j) \right\}.$$

Then

$$\begin{bmatrix} \xi_{11} & \xi_{21} \\ \xi_{12} & \xi_{22} \end{bmatrix} \in \mathrm{Reg}(A, M)$$

if and only if for all $\mu_{jt} \in \mathrm{Hom}_R(M_j, M_t)$ and for all $\alpha_{si} \in \mathrm{Hom}_R(A_s, A_i)$

$$\mu_{jt} \xi_{ij} \alpha_{si} \in \mathrm{Reg}(A_s, M_t).$$

Proof. Suppose that $\begin{bmatrix} \xi_{11} & \xi_{21} \\ \xi_{12} & \xi_{22} \end{bmatrix} \in \mathrm{Reg}(A, M)$. Recall that $\mathrm{Reg}(A, M)$ is an S-T-bimodule. Hence

$$\begin{bmatrix} \mu_{11} & 0 \\ 0 & 0 \end{bmatrix} \begin{bmatrix} \xi_{11} & \xi_{21} \\ \xi_{12} & \xi_{22} \end{bmatrix} \begin{bmatrix} \alpha_{11} & 0 \\ 0 & 0 \end{bmatrix} = \begin{bmatrix} \mu_{11}\xi_{11}\alpha_{11} & 0 \\ 0 & 0 \end{bmatrix} \in \mathrm{Reg}(A, M).$$

By Theorem 6.11 it follows that $\mu_{11}\xi_{11}\alpha_{11} \in \mathrm{Hom}_R(A_1, M_1)$ is regular. Similarly,

$$\begin{bmatrix} 0 & \mu_{21} \\ 0 & 0 \end{bmatrix} \begin{bmatrix} \xi_{11} & \xi_{21} \\ \xi_{12} & \xi_{22} \end{bmatrix} \begin{bmatrix} 0 & 0 \\ \alpha_{12} & 0 \end{bmatrix} = \begin{bmatrix} \mu_{21}\xi_{22}\alpha_{12} & 0 \\ 0 & 0 \end{bmatrix} \in \mathrm{Reg}(A, M).$$

By Theorem 6.11 it follows that $\mu_{21}\xi_{22}\alpha_{12} \in \operatorname{Hom}_R(A_1, M_1)$ is regular. Similarly,

$$\begin{bmatrix} 0 & 0 \\ 0 & \mu_{22} \end{bmatrix} \begin{bmatrix} \xi_{11} & \xi_{21} \\ \xi_{12} & \xi_{22} \end{bmatrix} \begin{bmatrix} \alpha_{11} & 0 \\ 0, & 0 \end{bmatrix} = \begin{bmatrix} 0 & 0 \\ \mu_{22}\xi_{12}\alpha_{11} & 0 \end{bmatrix} \in \operatorname{Reg}(A, M).$$

By Theorem 6.11 it follows that $\mu_{22}\xi_{12}\alpha_{11} \in \operatorname{Hom}_R(A_1, M_2)$ is regular.

The remaining cases work mutatis mutandis (see the Appendix to this section for the relevant matrix identities).

We have shown that $\mu(\mu_{jt}\xi_{ij}\alpha_{si})\alpha = (\mu\mu_{jt})\xi_{ij}(\alpha_{si}\alpha)$ is regular for all $\mu \in \operatorname{End}(M_t)$ and all $\alpha \in \operatorname{End}(A_s)$. It is left to show that sums of such elements are regular. Consider $\sum_k \mu_{jt}^{(k)}\xi_{ij}\alpha_{si}^{(k)}$. Using suitable formulas in the appendix we have matrices

$$\begin{bmatrix} 0 & \mu_{jt}^{(k)}\xi_{ij}\alpha_{si}^{(k)} \\ 0 & 0 \end{bmatrix} \in \operatorname{Reg}(A, M)$$

where the nonzero entry is in position (t, s) in the matrix (the displayed case is $s = 2$, $t = 1$). Hence

$$\sum_k \begin{bmatrix} 0 & \mu_{jt}^{(k)}\xi_{ij}\alpha_{si}^{(k)} \\ 0 & 0 \end{bmatrix} = \begin{bmatrix} 0 & \sum_k \mu_{jt}^{(k)}\xi_{ij}\alpha_{si}^{(k)} \\ 0 & 0 \end{bmatrix} \in \operatorname{Reg}(A, M);$$

it follows that $\sum_k \mu_{jt}^{(k)}\xi_{ij}\alpha_{si}^{(k)} \in \operatorname{Reg}(A_s, M_t)$. Hence $\operatorname{End}(M_t)(\mu_{jt}\xi_{ij}\alpha_{si})\operatorname{End}(A_s)$ is regular and therefore $\mu_{jt}\xi_{ij}\alpha_{si} \in \operatorname{Reg}(A_s, M_t)$.

Conversely, let $\xi = \begin{bmatrix} \xi_{11} & \xi_{21} \\ \xi_{12} & \xi_{22} \end{bmatrix} \in \operatorname{Hom}_R(A, M)$ and assume that for all $\mu_{jt} \in \operatorname{Hom}_R(M_j, M_t)$ and for all $\alpha_{si} \in \operatorname{Hom}_R(A_s, A_i)$,

$$\mu_{jt}\xi_{ij}\alpha_{si} \in \operatorname{Reg}(A_s, M_t).$$

Assume first that $\xi_{12} = 0$. By assumption there is $\eta_{11} \in \operatorname{Hom}_R(M_1, A_1)$ such that $\xi_{11}\eta_{11}\xi_{11} = \xi_{11}$ and there is $\eta_{22} \in \operatorname{Hom}_R(M_2, A_2)$ such that $\xi_{22}\eta_{22}\xi_{22} = \xi_{22}$. Then

$$\begin{aligned} \xi \begin{bmatrix} \eta_{11} & 0 \\ 0 & \eta_{22} \end{bmatrix} \xi &= \begin{bmatrix} \xi_{11} & \xi_{21} \\ 0 & \xi_{22} \end{bmatrix} \begin{bmatrix} \eta_{11} & 0 \\ 0 & \eta_{22} \end{bmatrix} \begin{bmatrix} \xi_{11} & \xi_{21} \\ 0 & \xi_{22} \end{bmatrix} \\ &= \begin{bmatrix} \xi_{11}\eta_{11}\xi_{11} & \xi_{11}\eta_{11}\xi_{21} + \xi_{21}\eta_{22}\xi_{22} \\ 0 & \xi_{22}\eta_{22}\xi_{22} \end{bmatrix} \\ &= \xi + \begin{bmatrix} 0 & \xi_{11}\eta_{11}\xi_{21} + \xi_{21}\eta_{22}\xi_{22} - \xi_{21} \\ 0 & 0 \end{bmatrix} \\ &= \xi + \begin{bmatrix} 0 & \xi'_{21} \\ 0 & 0 \end{bmatrix} \end{aligned}$$

where $\xi'_{21} = \xi_{11}\eta_{11}\xi_{21} + \xi_{21}\eta_{22}\xi_{22} - \xi_{21} \in \operatorname{Reg}(A_2, M_1)$ by hypothesis. Hence there is $\eta_{12} \in \operatorname{Hom}_R(M_1, A_2)$ such that

$$\xi'_{21}\eta_{12}\xi'_{21} = \xi'_{21}.$$

Then

$$\xi \begin{bmatrix} \eta_{11} & 0 \\ 0 & \eta_{22} \end{bmatrix} \xi = \xi + \begin{bmatrix} 0 & \xi'_{21} \\ 0 & 0 \end{bmatrix} = \xi + \begin{bmatrix} 0 & \xi'_{21} \\ 0 & 0 \end{bmatrix} \begin{bmatrix} 0 & 0 \\ \eta_{12} & 0 \end{bmatrix} \begin{bmatrix} 0 & \xi'_{21} \\ 0 & 0 \end{bmatrix}$$

$$= \xi + \left(\xi \begin{bmatrix} \eta_{11} & 0 \\ 0 & \eta_{22} \end{bmatrix} \xi - \xi \right) \begin{bmatrix} 0 & 0 \\ \eta_{12} & 0 \end{bmatrix} \left(\xi \begin{bmatrix} \eta_{11} & 0 \\ 0 & \eta_{22} \end{bmatrix} \xi - \xi \right)$$

$$= \xi + \xi \left(\begin{bmatrix} \eta_{11} & 0 \\ 0 & \eta_{22} \end{bmatrix} \xi - 1 \right) \begin{bmatrix} 0 & 0 \\ \eta_{12} & 0 \end{bmatrix} \left(\xi \begin{bmatrix} \eta_{11} & 0 \\ 0 & \eta_{22} \end{bmatrix} - 1 \right) \xi.$$

Using that $\xi X \xi = \xi + \xi Y \xi$ if and only if $\xi(X - Y)\xi = \xi$, this shows that ξ is regular.

Now consider a general element $\xi = \begin{bmatrix} \xi_{11} & \xi_{21} \\ \xi_{12} & \xi_{22} \end{bmatrix} \in \mathrm{Hom}_R(A, M)$ such that for all $\mu_{jt} \in \mathrm{Hom}_R(M_j, M_t)$ and for all $\alpha_{si} \in \mathrm{Hom}_R(A_s, A_i)$ it is true that $\mu_{jt} \xi_{ij} \alpha_{si} \in \mathrm{Reg}(A_s, M_t)$.

Choose $\eta_{21} \in \mathrm{Hom}_R(M_2, A_1)$ such that $\xi_{12} \eta_{21} \xi_{12} = \xi_{12}$. Then

$$\xi \begin{bmatrix} 0 & \eta_{21} \\ 0 & 0 \end{bmatrix} \xi = \begin{bmatrix} \xi_{11} & \xi_{21} \\ \xi_{12} & \xi_{22} \end{bmatrix} \begin{bmatrix} 0 & \eta_{21} \\ 0 & 0 \end{bmatrix} \begin{bmatrix} \xi_{11} & \xi_{21} \\ \xi_{12} & \xi_{22} \end{bmatrix}$$

$$= \begin{bmatrix} 0 & \xi_{11}\eta_{21} \\ 0 & \xi_{12}\eta_{21} \end{bmatrix} \begin{bmatrix} \xi_{11} & \xi_{21} \\ \xi_{12} & \xi_{22} \end{bmatrix}$$

$$= \begin{bmatrix} \xi_{11}\eta_{21}\xi_{12} & \xi_{11}\eta_{21}\xi_{22} \\ \xi_{12}\eta_{21}\xi_{12} & \xi_{12}\eta_{21}\xi_{22} \end{bmatrix} = \begin{bmatrix} \xi_{11}\eta_{21}\xi_{12} & \xi_{11}\eta_{21}\xi_{22} \\ \xi_{12} & \xi_{12}\eta_{21}\xi_{22} \end{bmatrix}.$$

Hence

$$\xi \begin{bmatrix} 0 & \eta_{21} \\ 0 & 0 \end{bmatrix} \xi - \xi = \begin{bmatrix} \xi_{11}\eta_{21}\xi_{12} - \xi_{11} & \xi_{11}\eta_{21}\xi_{22} - \xi_{21} \\ 0 & \xi_{12}\eta_{21}\xi_{22} - \xi_{22} \end{bmatrix}$$

which is regular by the special case treated first. It follows that ξ is regular. We need to show further that every element in $\mathrm{End}(M) \xi \, \mathrm{End}(A)$ is regular. Let

$$\alpha := \begin{bmatrix} \alpha_{11} & \alpha_{21} \\ \alpha_{12} & \alpha_{22} \end{bmatrix} \in \mathrm{End}(A) \quad \text{and} \quad \mu := \begin{bmatrix} \mu_{11} & \mu_{21} \\ \mu_{12} & \mu_{22} \end{bmatrix} \in \mathrm{End}(M).$$

Then

$$\mu\xi\alpha = \begin{bmatrix} \mu_{11}\xi_{11}\alpha_{11} + \mu_{11}\xi_{21}\alpha_{12} + \mu_{21}\xi_{12}\alpha_{11} + \mu_{21}\xi_{22}\alpha_{12} \\ \mu_{11}\xi_{11}\alpha_{21} + \mu_{11}\xi_{21}\alpha_{22} + \mu_{21}\xi_{12}\alpha_{21} + \mu_{21}\xi_{22}\alpha_{22} \\ \mu_{12}\xi_{11}\alpha_{11} + \mu_{12}\xi_{21}\alpha_{12} + \mu_{22}\xi_{12}\alpha_{11} + \mu_{22}\alpha_{12}\xi_{22} \\ \mu_{12}\xi_{11}\alpha_{21} + \mu_{12}\xi_{21}\alpha_{22} + \mu_{22}\xi_{12}\alpha_{21} + \mu_{22}\xi_{22}\alpha_{22} \end{bmatrix}.$$

Then for $\mu'_{js} \in \mathrm{Hom}_R(M_j, M_s)$ and $\alpha'_{ti} \in \mathrm{Hom}_R(A_t, A_i)$ we have

$$\mu'_{1s} (\mu_{11}\xi_{11}\alpha_{11} + \mu_{11}\xi_{21}\alpha_{12} + \mu_{21}\xi_{12}\alpha_{11} + \mu_{21}\xi_{22}\alpha_{12}) \alpha'_{t1}$$

$$= (\mu'_{1s}\mu_{11})\xi_{11}(\alpha_{11}\alpha'_{t1}) + (\mu'_{1s}\mu_{11})\xi_{21}(\alpha_{12}\alpha'_{t1})$$

$$+ (\mu'_{1s}\mu_{21})\xi_{12}(\alpha_{11}\alpha'_{t1}) + (\mu'_{1s}\mu_{21})\xi_{22}(\alpha_{12}\alpha'_{t1})$$

$$\in \mathrm{Reg}(A_t, M_s),$$

$$\mu'_{1s}\left(\mu_{11}\xi_{11}\alpha_{21} + \mu_{11}\xi_{21}\alpha_{22} + \mu_{21}\xi_{12}\alpha_{21} + \mu_{21}\xi_{22}\alpha_{22}\right)\alpha'_{t2}$$
$$= (\mu'_{1s}\mu_{11})\xi_{11}(\alpha_{21}\alpha'_{t2}) + (\mu'_{1s}\mu_{11})\xi_{21}(\alpha_{22}\alpha'_{t2})$$
$$\quad + (\mu'_{1s}\mu_{21})\xi_{12}(\alpha_{21}\alpha'_{t2}) + (\mu'_{1s}\mu_{21})\xi_{22}(\alpha_{22}\alpha'_{t2})$$
$$\in \mathrm{Reg}(A_t, M_s),$$

$$\mu'_{2s}\left(\mu_{12}\xi_{11}\alpha_{11} + \mu_{12}\xi_{21}\alpha_{12} + \mu_{22}\xi_{12}\alpha_{11} + \mu_{22}\xi_{22}\alpha_{12}\right)\alpha'_{t1}$$
$$= (\mu'_{2s}\mu_{12})\xi_{11}(\alpha_{11}\alpha'_{t1}) + (\mu'_{2s}\mu_{12})\xi_{21}(\alpha_{12}\alpha'_{t1})$$
$$\quad + (\mu'_{2s}\mu_{22})\xi_{12}(\alpha_{11}\alpha'_{t1}) + (\mu'_{2s}\mu_{22})\xi_{22}(\alpha_{12}\alpha'_{t1})$$
$$\in \mathrm{Reg}(A_t, M_s),$$

$$\mu'_{2s}\left(\mu_{12}\xi_{11}\alpha_{21} + \mu_{12}\xi_{21}\alpha_{22} + \mu_{22}\xi_{12}\alpha_{21} + \mu_{22}\xi_{22}\alpha_{22}\right)\alpha'_{t2}$$
$$= (\mu'_{2s}\mu_{12})\xi_{11}(\alpha_{21}\alpha'_{t2}) + (\mu'_{2s}\mu_{12})\xi_{21}(\alpha_{22}\alpha'_{t2})$$
$$\quad + (\mu'_{2s}\mu_{22})\xi_{12}(\alpha_{21}\alpha'_{t2}) + (\mu'_{2s}\mu_{22})\xi_{22}(\alpha_{22}\alpha'_{t2})$$
$$\in \mathrm{Reg}(A_t, M_s).$$

By what has been shown, $\mu\xi\alpha$ is regular in $\mathrm{Hom}_R(A, M)$. Finally, we must show that the sums $\sum_k \mu^{(k)}\xi\alpha^{(k)}$ are regular. The entry in position (s,t) of a matrix $\mu^{(k)}\xi\alpha^{(k)}$ are sums of elements of the form $\mu^{(k)}_{jt}\xi_{ij}\alpha^{(k)}_{si}$ and so are the entries of the matrix $\sum_k \mu^{(k)}\xi\alpha^{(k)}$. These entries are in $\mathrm{Reg}(A_s, M_t)$ and hence $\sum_k \mu^{(k)}\xi\alpha^{(k)} \in \mathrm{Reg}(A, M)$. This completes the proof that $\xi \in \mathrm{Reg}(A, M)$. $\square$

Using the characterization of regular maps we have a useful consequence.

Corollary 6.15. *Let $A = A_1 \oplus A_2$ and $M = M_1 \oplus M_2$ as before. Suppose that*

$$\begin{bmatrix} \xi_{11} & \xi_{21} \\ \xi_{12} & \xi_{22} \end{bmatrix} \in \mathrm{Reg}(A, M).$$

Then for all $\mu_{jt} \in \mathrm{Hom}_R(M_j, M_t)$ and for all $\alpha_{si} \in \mathrm{Hom}_R(A_s, A_i)$,

$$\mathrm{Ker}(\mu_{jt}\xi_{ij}\alpha_{si}) \subseteq^{\oplus} A_s \quad and \quad \mathrm{Im}(\mu_{jt}\xi_{ij}\alpha_{si}) \subseteq^{\oplus} M_t.$$

We will use later the following special case.

Corollary 6.16. *Let A and M be modules such that $A = A_1 \oplus A_2$ and $M = M_1 \oplus M_2$ and no nonzero summand of A_2 is isomorphic to a summand of M_1 and no nonzero summand of A_1 is isomorphic to a summand of M_2. Suppose that*

$$\begin{bmatrix} \xi_{11} & \xi_{21} \\ \xi_{12} & \xi_{22} \end{bmatrix} \in \mathrm{Reg}(A, M).$$

Then $\xi_{21}=0$, $\xi_{12}=0$, and for every $\mu_{21} \in \mathrm{Hom}(M_2, M_1)$, every $\alpha_{21} \in \mathrm{Hom}(A_2, A_1)$, every $\mu_{12} \in \mathrm{Hom}(M_1, M_2)$, and every $\alpha_{12} \in \mathrm{Hom}(A_1, A_2)$,

$$\mu_{21}\xi_{22} = 0, \quad \xi_{11}\alpha_{21} = 0, \quad \mu_{12}\xi_{11} = 0, \quad \xi_{22}\alpha_{12} = 0,$$

and

$$\mu_{21}\xi_{22}\alpha_{12} \in \mathrm{Reg}(A_1, M_1), \quad \mu_{12}\xi_{11}\alpha_{21} \in \mathrm{Reg}(A_2, M_2).$$

Proof. By hypothesis and Corollary 4.4, $\mathrm{Reg}(A_1, M_2) = 0$ and $\mathrm{Reg}(A_2, M_1) = 0$. By Theorem 6.14, $\xi_{21} = 0$ and $\xi_{12} = 0$. We also have that for all $\begin{bmatrix} \mu_{11} & \mu_{21} \\ \mu_{12} & \mu_{22} \end{bmatrix} \in$ $\mathrm{End}(M)$ and all $\begin{bmatrix} \alpha_{11} & \alpha_{21} \\ \alpha_{12} & \alpha_{22} \end{bmatrix} \in \mathrm{End}(A)$,

$$\begin{bmatrix} \mu_{11}\xi_{11}\alpha_{11} + \mu_{21}\xi_{22}\alpha_{12} & \mu_{11}\xi_{11}\alpha_{21} + \mu_{21}\xi_{22}\alpha_{22} \\ \mu_{12}\xi_{11}\alpha_{11} + \mu_{22}\xi_{22}\alpha_{12} & \mu_{12}\xi_{11}\alpha_{21} + \mu_{22}\xi_{22}\alpha_{22} \end{bmatrix}$$
$$= \begin{bmatrix} \mu_{11} & \mu_{21} \\ \mu_{12} & \mu_{22} \end{bmatrix} \begin{bmatrix} \xi_{11} & 0 \\ 0 & \xi_{22} \end{bmatrix} \begin{bmatrix} \alpha_{11} & \alpha_{21} \\ \alpha_{12} & \alpha_{22} \end{bmatrix} \in \mathrm{Reg}(A, M).$$

By Theorem 6.14 it follows that $\mu_{11}\xi_{11}\alpha_{21} + \mu_{21}\xi_{22}\alpha_{22} \in \mathrm{Reg}(A_2, M_1) = 0$. Choosing $\mu_{11} = 0$ and $\alpha_{22} = 1$, we find that $\mu_{21}\xi_{22} = 0$. The analogous claims follow similarly. Finally, by Theorem 6.14 we get $\mu_{11}\xi_{11}\alpha_{11} \in \mathrm{Reg}(A_1, M_1)$ and $\mu_{12}\xi_{11}\alpha_{21} \in \mathrm{Reg}(A_2, M_2)$. $\qquad\square$

Theorem 6.14 can be extended to arbitrary finite sums by induction.

Theorem 6.17. *Let* $A = A_1 \oplus \cdots \oplus A_m$ *and* $M = M_1 \oplus \cdots \oplus M_n$. *Then*

$$\xi = \begin{bmatrix} \xi_{11} & \cdots & \xi_{m1} \\ \vdots & \vdots & \vdots \\ \xi_{1n} & \cdots & \xi_{mn} \end{bmatrix} \in \mathrm{Reg}(A, M)$$

if and only if for all $\mu_{jt} \in \mathrm{Hom}_R(M_j, M_t)$ *and for all* $\alpha_{si} \in \mathrm{Hom}_R(A_s, A_i)$,

$$\mu_{jt}\xi_{ij}\alpha_{si} \in \mathrm{Reg}(A_s, M_t).$$

Proof. Set $A'_1 := A_1$, $A'_2 := A_2 \oplus \cdots \oplus A_m$, $M'_1 := M_1$, and $M'_2 := M_2 \oplus \cdots \oplus M_n$. Then $A = A'_1 \oplus A'_2$ and $M = M'_1 \oplus M'_n$. Then

$$\xi = \begin{bmatrix} \eta_{11} & \eta_{21} \\ \eta_{12} & \eta_{22} \end{bmatrix}$$

where

$$\eta_{11} = \xi_{11}, \ \eta_{21} = \begin{bmatrix} \xi_{21} & \cdots & \xi_{n1} \end{bmatrix}, \ \eta_{12} = \begin{bmatrix} \xi_{12} \\ \vdots \\ \xi_{1m} \end{bmatrix}, \ \eta_{22} = \begin{bmatrix} \xi_{22} & \cdots & \xi_{n2} \\ \vdots & \vdots & \vdots \\ \xi_{2m} & \cdots & \xi_{nm} \end{bmatrix}.$$

We now wish to apply induction using Theorem 6.14. To do so we must consider homomorphisms induced by ξ on the subsums A_1 and A'_2 of A to the subsums M_1 and M'_2 of M. All such homomorphisms are matrices with entries $\xi_{ij} \in \mathrm{Hom}_R(A_i, M_j)$.

We must then check whether the conditions of Theorem 6.14 prevail. To do so we must apply the endomorphisms of the subsums. All such endomorphisms are matrices with entries $\mu_{ij} \in \mathrm{Hom}_R(M_i, M_j)$ and $\alpha_{ij} \in \mathrm{Hom}_R(A_i, A_j)$ for various sets of subscripts. Looking at products $\mu\eta\alpha$ we see that their entries are of the form $\sum_{(i,j)} \mu_{js}\xi_{ij}\alpha_{ti} \in \mathrm{Hom}_R(A_t, M_s)$. In one direction we have that each summand is in some Reg, hence also the sum, and in the other direction we know that the sum is in some Reg for all choices of the μ_{js} and α_{ti} and choosing all but one of the μ and all but one of the α to be 0, we get the necessity of the conditions $\mu_{js}\xi_{ij}\alpha_{ti} \in \mathrm{Reg}(A_t, M_s)$. $\square$

Corollary 6.18. *Let $A = A_1 \oplus \cdots \oplus A_m$ and $M = M_1 \oplus \cdots \oplus M_n$. Then $\mathrm{Hom}_R(A, M)$ is regular, i.e., $\mathrm{Reg}(A, M) = \mathrm{Hom}_R(A, M)$, if and only if*

$$\forall i, j : \mathrm{Hom}(A_i, M_j) = \mathrm{Reg}(A_i, M_j).$$

Corollary 6.19. *Let $A = A_1 \oplus \cdots \oplus A_m$, $M = M_1 \oplus \cdots \oplus M_n$, and N be R-modules such that $A_i \cong N \cong M_j$ for all i and j. Then the following statements hold.*

1) $\mathrm{Reg}(A, M) \neq 0$ *if and only if* $\mathrm{Reg}(N, N) \neq 0$.

2) $\mathrm{Reg}(A, M) = \mathrm{Hom}_R(A, M)$ *if and only if* $\mathrm{Reg}(N, N) = \mathrm{End}(N_R)$.

3) $\mathrm{End}(N^k)$ *is regular if and only if* $\mathrm{End}(N)$ *is regular.*

Proof. Select isomorphisms $\sigma_i : A_i \to N$ and $\tau_j : M_j \to N$.

1) Suppose that $\mathrm{Reg}(N, N) \neq 0$ and let $0 \neq \eta \in \mathrm{Reg}(N, N)$. Then $\tau_1^{-1}\eta\sigma_1 \in \mathrm{Hom}_R(A_1, M_1)$ and $\tau_1^{-1}\eta\sigma_1 \neq 0$. Let

$$\xi := \begin{bmatrix} \tau_1^{-1}\eta\sigma_1 & 0 & \cdots & 0 \\ 0 & 0 & \cdots & 0 \\ \vdots & \vdots & \cdots & \vdots \\ 0 & 0 & \cdots & 0 \end{bmatrix} \in \mathrm{Hom}_R(A, M).$$

By Theorem 6.17, $0 \neq \xi \in \mathrm{Reg}(A, M)$ if and only if for all $\alpha_{i1} \in \mathrm{Hom}_R(A_i, A_1)$ and all μ_{1j} it is true that $\mu_{1j}(\tau_1^{-1}\eta\sigma_1)\alpha_{i1} \in \mathrm{Reg}(A_i, M_j)$. But $\tau_j(\mu_{1j}\tau_1^{-1}\eta\sigma_1\alpha_{i1})\sigma_i^{-1} = (\tau_j\mu_{1j}\tau_1^{-1})\eta(\sigma_1\alpha_{i1}\sigma_i^{-1}) \in (\mathrm{End}\,N)\eta(\mathrm{End}\,N) \subseteq \mathrm{Reg}(N, N)$ and it follows that $\mu_{1j}(\tau_1^{-1}\eta\sigma_1)\alpha_{i1} \in \mathrm{Reg}(A_i, M_j)$.

Conversely, suppose that $\mathrm{Reg}(A, M) \neq 0$ and let

$$0 \neq \xi = \begin{bmatrix} \xi_{11} & \cdots & \xi_{m1} \\ \vdots & \vdots & \vdots \\ \xi_{1n} & \cdots & \xi_{mn} \end{bmatrix} \in \mathrm{Reg}(A, M).$$

Then there is an entry $\xi_{ij} \neq 0$ in position (j, i). Let $E_{j1} \in \mathrm{End}(M)$ be the $n \times n$-matrix with entry $\tau_1\tau_j^{-1}$ in position $(1, j)$ and 0 everywhere else, and let $E_{1i} \in$

End(A) be the $m \times m$ matrix with entry $\sigma_i^{-1}\sigma_1$ in position $(i, 1)$ and 0 everywhere else. Then

$$0 \neq E_{j1}\xi E_{1i} = \begin{bmatrix} \tau_1\tau_j^{-1}\xi_{ij}\sigma_i\sigma_1^{-1} & \cdots & 0 \\ \vdots & \vdots & \vdots \\ 0 & \cdots & 0 \end{bmatrix} \in \mathrm{Reg}(A, M).$$

It is now clear that $0 \neq \tau_1\tau_i^{-1}\xi_{ji}\sigma_j\sigma_1^{-1} \in \mathrm{Reg}(A_1, M_1)$. So $\mathrm{Reg}(A_1, M_1) \neq 0$, and hence also $\mathrm{Reg}(N, N) \neq 0$.

2) is an immediate consequence of Theorem 6.17 and 3) is a special case of 2). $\qquad\square$

Corollary 6.20. *The matrix ring* $\mathbb{M}_n(R)$ *is regular if and only if* $\mathrm{End}(R_R)$ *is regular.*

7 Appendix: Various Formulas

$$\begin{bmatrix} \mu_{11} & \mu_{21} \\ \mu_{12} & \mu_{22} \end{bmatrix} \begin{bmatrix} \xi_{11} & \xi_{21} \\ \xi_{12} & \xi_{22} \end{bmatrix} \begin{bmatrix} \alpha_{11} & \alpha_{21} \\ \alpha_{12} & \alpha_{22} \end{bmatrix}$$

$$= \begin{bmatrix} \mu_{11} & \mu_{21} \\ \mu_{12} & \mu_{22} \end{bmatrix} \begin{bmatrix} \xi_{11}\alpha_{11} + \xi_{21}\alpha_{12} & \xi_{11}\alpha_{21} + \xi_{21}\alpha_{22} \\ \xi_{12}\alpha_{11} + \xi_{22}\alpha_{12} & \xi_{12}\alpha_{21} + \alpha_{22}\xi_{22} \end{bmatrix}$$

$$= \begin{bmatrix} \mu_{11}\xi_{11}\alpha_{11} + \mu_{11}\xi_{21}\alpha_{12} + \mu_{21}\xi_{12}\alpha_{11} + \mu_{21}\xi_{22}\alpha_{12} \\ \mu_{11}\xi_{11}\alpha_{21} + \mu_{11}\xi_{21}\alpha_{22} + \mu_{21}\xi_{12}\alpha_{21} + \mu_{21}\xi_{22}\alpha_{22} \\ \mu_{12}\xi_{11}\alpha_{11} + \mu_{12}\xi_{21}\alpha_{12} + \mu_{22}\xi_{12}\alpha_{11} + \mu_{22}\xi_{22}\alpha_{12} \\ \mu_{12}\xi_{11}\alpha_{21} + \mu_{12}\xi_{21}\alpha_{22} + \mu_{22}\xi_{12}\alpha_{21} + \mu_{22}\xi_{22}\alpha_{22} \end{bmatrix}$$

In general,

$$[\mu_{ij}][\xi_{ij}][\alpha_{ij}] = \left[\sum_{(i,j)} \mu_{js}\xi_{ij}\alpha_{ti}\right].$$

There are 16 possibilities for the maps $\mu_{jt}\xi_{ij}\alpha_{si}$ namely:

$$\left.\begin{array}{l} \mu_{11}\xi_{11}\alpha_{11} \\ \mu_{21}\xi_{12}\alpha_{11} \\ \mu_{11}\xi_{21}\alpha_{12} \\ \mu_{21}\xi_{22}\alpha_{12} \end{array}\right\} \in \mathrm{Hom}_R(A_1, M_1), \qquad \left.\begin{array}{l} \mu_{12}\xi_{11}\alpha_{11} \\ \mu_{22}\xi_{12}\alpha_{11} \\ \mu_{12}\xi_{21}\alpha_{12} \\ \mu_{22}\xi_{22}\alpha_{12} \end{array}\right\} \in \mathrm{Hom}_R(A_1, M_2),$$

$$\left.\begin{array}{l} \mu_{11}\xi_{11}\alpha_{21} \\ \mu_{21}\xi_{12}\alpha_{21} \\ \mu_{11}\xi_{21}\alpha_{22} \\ \mu_{21}\xi_{22}\alpha_{22} \end{array}\right\} \in \mathrm{Hom}_R(A_2, M_1), \qquad \left.\begin{array}{l} \mu_{12}\xi_{11}\alpha_{21} \\ \mu_{22}\xi_{12}\alpha_{21} \\ \mu_{12}\xi_{21}\alpha_{22} \\ \mu_{22}\xi_{22}\alpha_{22} \end{array}\right\} \in \mathrm{Hom}_R(A_2, M_2),$$

and there are 16 corresponding matrix equations:

$$
\begin{bmatrix} \mu_{11}\xi_{11}\alpha_{11} & 0 \\ 0 & 0 \end{bmatrix} = \begin{bmatrix} \mu_{11} & 0 \\ 0 & 0 \end{bmatrix} \begin{bmatrix} \xi_{11} & \xi_{21} \\ \xi_{12} & \xi_{22} \end{bmatrix} \begin{bmatrix} \alpha_{11} & 0 \\ 0 & 0 \end{bmatrix} \tag{23}
$$

$$
\begin{bmatrix} \mu_{21}\xi_{12}\alpha_{11} & 0 \\ 0 & 0 \end{bmatrix} = \begin{bmatrix} 0 & \mu_{21} \\ 0 & 0 \end{bmatrix} \begin{bmatrix} \xi_{11} & \xi_{21} \\ \xi_{12} & \xi_{22} \end{bmatrix} \begin{bmatrix} \alpha_{11} & 0 \\ 0 & 0 \end{bmatrix} \tag{24}
$$

$$
\begin{bmatrix} \mu_{11}\xi_{21}\alpha_{12} & 0 \\ 0 & 0 \end{bmatrix} = \begin{bmatrix} \mu_{11} & 0 \\ 0 & 0 \end{bmatrix} \begin{bmatrix} \xi_{11} & \xi_{21} \\ \xi_{12} & \xi_{22} \end{bmatrix} \begin{bmatrix} 0 & 0 \\ \alpha_{12} & 0 \end{bmatrix} \tag{25}
$$

$$
\begin{bmatrix} \mu_{21}\xi_{22}\alpha_{12} & 0 \\ 0 & 0 \end{bmatrix} = \begin{bmatrix} 0 & \mu_{21} \\ 0 & 0 \end{bmatrix} \begin{bmatrix} \xi_{11} & \xi_{21} \\ \xi_{12} & \xi_{22} \end{bmatrix} \begin{bmatrix} 0 & 0 \\ \alpha_{12} & 0 \end{bmatrix} \tag{26}
$$

$$
\begin{bmatrix} 0 & 0 \\ \mu_{12}\xi_{11}\alpha_{11} & 0 \end{bmatrix} = \begin{bmatrix} 0 & 0 \\ \mu_{12} & 0 \end{bmatrix} \begin{bmatrix} \xi_{11} & \xi_{21} \\ \xi_{12} & \xi_{22} \end{bmatrix} \begin{bmatrix} \alpha_{11} & 0 \\ 0 & 0 \end{bmatrix} \tag{27}
$$

$$
\begin{bmatrix} 0 & 0 \\ \mu_{22}\xi_{12}\alpha_{11} & 0 \end{bmatrix} = \begin{bmatrix} 0 & 0 \\ 0 & \mu_{22} \end{bmatrix} \begin{bmatrix} \xi_{11} & \xi_{21} \\ \xi_{12} & \xi_{22} \end{bmatrix} \begin{bmatrix} \alpha_{11} & 0 \\ 0 & 0 \end{bmatrix} \tag{28}
$$

$$
\begin{bmatrix} 0 & 0 \\ \mu_{12}\xi_{21}\alpha_{12} & 0 \end{bmatrix} = \begin{bmatrix} 0 & 0 \\ \mu_{12} & 0 \end{bmatrix} \begin{bmatrix} \xi_{11} & \xi_{21} \\ \xi_{12} & \xi_{22} \end{bmatrix} \begin{bmatrix} 0 & 0 \\ \alpha_{12} & 0 \end{bmatrix} \tag{29}
$$

$$
\begin{bmatrix} 0 & 0 \\ \mu_{22}\xi_{22}\alpha_{12} & 0 \end{bmatrix} = \begin{bmatrix} 0 & 0 \\ 0 & \mu_{22} \end{bmatrix} \begin{bmatrix} \xi_{11} & \xi_{21} \\ \xi_{12} & \xi_{22} \end{bmatrix} \begin{bmatrix} 0 & 0 \\ \alpha_{12} & 0 \end{bmatrix} \tag{30}
$$

$$
\begin{bmatrix} 0 & \mu_{11}\xi_{11}\alpha_{21} \\ 0 & 0 \end{bmatrix} = \begin{bmatrix} \mu_{11} & 0 \\ 0 & 0 \end{bmatrix} \begin{bmatrix} \xi_{11} & \xi_{21} \\ \xi_{12} & \xi_{22} \end{bmatrix} \begin{bmatrix} 0 & \alpha_{21} \\ 0 & 0 \end{bmatrix} \tag{31}
$$

$$
\begin{bmatrix} 0 & \mu_{21}\xi_{12}\alpha_{21} \\ 0 & 0 \end{bmatrix} = \begin{bmatrix} 0 & \mu_{21} \\ 0 & 0 \end{bmatrix} \begin{bmatrix} \xi_{11} & \xi_{21} \\ \xi_{12} & \xi_{22} \end{bmatrix} \begin{bmatrix} 0 & \alpha_{21} \\ 0 & 0 \end{bmatrix} \tag{32}
$$

$$
\begin{bmatrix} 0 & \mu_{11}\xi_{21}\alpha_{22} \\ 0 & 0 \end{bmatrix} = \begin{bmatrix} \mu_{11} & 0 \\ 0 & 0 \end{bmatrix} \begin{bmatrix} \xi_{11} & \xi_{21} \\ \xi_{12} & \xi_{22} \end{bmatrix} \begin{bmatrix} 0 & 0 \\ 0 & \alpha_{22} \end{bmatrix} \tag{33}
$$

$$
\begin{bmatrix} 0 & \mu_{22}\xi_{22}\alpha_{22} \\ 0 & 0 \end{bmatrix} = \begin{bmatrix} 0 & \mu_{21} \\ 0 & 0 \end{bmatrix} \begin{bmatrix} \xi_{11} & \xi_{21} \\ \xi_{12} & \xi_{22} \end{bmatrix} \begin{bmatrix} 0 & 0 \\ 0 & \alpha_{22} \end{bmatrix} \tag{34}
$$

$$
\begin{bmatrix} 0 & 0 \\ 0 & \mu_{12}\xi_{11}\alpha_{21} \end{bmatrix} = \begin{bmatrix} 0 & 0 \\ \mu_{12} & 0 \end{bmatrix} \begin{bmatrix} \xi_{11} & \xi_{21} \\ \xi_{12} & \xi_{22} \end{bmatrix} \begin{bmatrix} 0 & \alpha_{21} \\ 0 & 0 \end{bmatrix} \tag{35}
$$

$$
\begin{bmatrix} 0 & 0 \\ 0 & \mu_{22}\xi_{12}\alpha_{21} \end{bmatrix} = \begin{bmatrix} 0 & 0 \\ 0 & \mu_{22} \end{bmatrix} \begin{bmatrix} \xi_{11} & \xi_{21} \\ \xi_{12} & \xi_{22} \end{bmatrix} \begin{bmatrix} 0 & \alpha_{21} \\ 0 & 0 \end{bmatrix} \tag{36}
$$

$$
\begin{bmatrix} 0 & 0 \\ 0 & \mu_{12}\xi_{21}\alpha_{22} \end{bmatrix} = \begin{bmatrix} 0 & 0 \\ \mu_{12} & 0 \end{bmatrix} \begin{bmatrix} \xi_{11} & \xi_{21} \\ \xi_{12} & \xi_{22} \end{bmatrix} \begin{bmatrix} 0 & 0 \\ 0 & \alpha_{22} \end{bmatrix} \tag{37}
$$

$$
\begin{bmatrix} 0 & 0 \\ 0 & \mu_{22}\xi_{22}\alpha_{22} \end{bmatrix} = \begin{bmatrix} 0 & 0 \\ 0 & \mu_{22} \end{bmatrix} \begin{bmatrix} \xi_{11} & \xi_{21} \\ \xi_{12} & \xi_{22} \end{bmatrix} \begin{bmatrix} 0 & 0 \\ 0 & \alpha_{22} \end{bmatrix} \tag{38}
$$

Chapter III

Indecomposable Modules

1 $\text{Reg}(A, M) \neq 0$

A module M is **directly indecomposable** (or simply **indecomposable**) if and only if 0 and M are the only direct summands of M. This means that 0 and 1 are the only idempotents in $\text{End}(M_R)$. We now study the situation that $\text{Reg}(A, M) \neq 0$ and one of the modules A or M is indecomposable. It turns out that much can be said under assumptions weaker than $\text{Reg}(A, M) \neq 0$.

Theorem 1.1.

1) *Suppose that there is $0 \neq f \in \text{Hom}_R(A, M)$ that is regular and μf is regular for all $\mu \in S := \text{End}(M_R)$. If M is directly indecomposable, then S is a division ring.*

2) *Suppose that there is $0 \neq f \in \text{Hom}_R(A, M)$ that is regular and $f\alpha$ is regular for all $\alpha \in T := \text{End}(A_R)$. If A is directly indecomposable, then T is a division ring.*

The proof of Theorem 1.1 requires a well-known lemma on division rings. For the sake of completeness we include it with proof.

Lemma 1.2. *Let R be a ring with $1 \in R$ and assume that every nonzero element of R has a right inverse. Then R is a division ring. The same is true for left inverses.*

Proof. Let $0 \neq r \in R$ and suppose that $rs = 1$ and $st = 1$. Then $t = 1t = rst = r1 = r$. Thus s is both right and left inverse of r and r is invertible. $\square$

Proof of Theorem 1.1. 1) Let M be directly indecomposable, and assume that $0 \neq f \in \text{Hom}_R(A, M)$ is regular. Hence $\text{Im}(f) \subseteq^\oplus M$ and M being indecomposable, it follows that $\text{Im}(f) = M$. Let $0 \neq \mu \in S$. Then $\mu f(A) = \mu M \neq 0$ and therefore $\mu f \neq 0$. By assumption also μf is regular. Hence there exists $g \in \text{Hom}_R(M, A)$ such that $\mu f g \mu f = \mu f$ and then $e := \mu f g = e^2 \neq 0$. Since M is indecomposable, its only nonzero idempotent is 1, so $e = 1$ and $\mu f g = \mu(fg) = 1$ with $fg \in S$, i.e., every $0 \neq \mu \in S$ possesses a right inverse. By Lemma 1.2 it follows that S is a division ring.

2) Let A be indecomposable, and assume that $0 \neq f \in \mathrm{Hom}_R(A, M)$ is regular. Then $\mathrm{Ker}(f) \subseteq^{\oplus} A$ and since $f \neq 0$ and A is indecomposable, $\mathrm{Ker}(f) = 0$. Let $0 \neq \alpha \in T$. Then there is $a \in A$ such that $\alpha(a) \neq 0$ and it follows that $f(\alpha(a)) \neq 0$, so $f\alpha \neq 0$. Since $f\alpha$ is regular by hypothesis, there exists $g \in \mathrm{Hom}_R(M, A)$ such that $f\alpha g f\alpha = f\alpha$. Hence $gf\alpha$ is a nonzero idempotent of T and we get $gf\alpha = 1$ and $gf \in T$ is a left inverse of α. By Lemma 1.2 it follows that T is a division ring.   □

If M is indecomposable and if $S := \mathrm{End}(M_R)$ is not a division ring then $\mathrm{Reg}(A, M) = 0$ for every $A \in \mathrm{Mod}\text{-}R$. Similarly, if A is indecomposable and $T := \mathrm{End}(A_R)$ is not a division ring then $\mathrm{Reg}(A, M) = 0$ for every $M \in \mathrm{Mod}\text{-}R$.

An example of a module M that is indecomposable but $\mathrm{End}(M_R)$ is not a division ring is $M = \mathbb{Z}/p^2\mathbb{Z}$ where p is a prime number.

Simple modules are examples of modules whose endomorphism rings are division rings. This is Schur's Lemma.

Example 1.3. The abelian groups whose endomorphism rings are division rings are the groups $\mathbb{Q}$ with $\mathrm{End}(\mathbb{Q}) = \mathbb{Q}$ and the cyclic groups A of order p with $\mathrm{End}(A) = \mathbb{Z}/p\mathbb{Z}$.

Theorem 1.4. *Let R be a commutative ring and M_R a faithful R-module. Suppose that $\mathrm{End}(M_R)$ is a division ring. Then R is an integral domain, $M = Q(R)$ and $\mathrm{End}(M_R) = Q(R)$ where $Q(R)$ is the field of quotients of R.*

Proof. Since R is commutative and M is faithful, $R \subseteq S := \mathrm{End}(M_R)$. Hence R is an integral domain and $Q(R) \subseteq S$. Hence M is a $Q(R)$-vector space. Let $Q(R)m$ be a one-dimensional subspace of M. Then $Q(R)m$ is an R-submodule of M and a direct summand. Since M is indecomposable, $M = Q(R)m \cong Q(R)$ and the result follows.   □

2   Structure Theorems

Suppose that $0 \neq f \in \mathrm{Reg}(A, M)$ and M is indecomposable. Then $A = \mathrm{Ker}(f) \oplus M'$ where $M' \cong M$ and $\mathrm{End}(M_R)$ is a division ring. Furthermore, for every $s \in S$ and $t \in T$, the map sft is regular (also sums of such). This being so, the module A must have a special structure. Similarly, if A is indecomposable and $\mathrm{Reg}(A, M) \neq 0$, then M must have a special structure.

We first settle the very special case when both M and A are indecomposable.

Proposition 2.1. *Suppose that A and M are indecomposable modules.*

1) *If $0 \neq f \in \mathrm{Hom}_R(A, M)$ is regular, then it is bijective and $A \cong M$.*

2) *If* $\text{Reg}(A, M) \neq 0$, *then* $A \cong M$ *and* $\text{End}(A_R) \cong \text{End}(M_R)$ *are division rings.*
If so, then $\text{Reg}(A, M) = \text{Hom}_R(A, M)$.

Proof. 1) We know that $A = \text{Ker}(f) \oplus \text{Im}(gf)$ and $M = \text{Im}(f) \oplus \text{Ker}(fg)$. As $f \neq 0$ and A and M are indecomposable, we have $\text{Ker}(f) = 0$ and $M = \text{Im}(f)$. Hence f is bijective.

2) By 1) $A \cong M$ and $\text{End}(A_R) \cong \text{End}(M_R)$ are division rings by Theorem 1.1. If so every nonzero element of $\text{Hom}_R(A, M)$ is a bijection and hence every map in $\text{Hom}_R(A, M)$ is regular. $\qquad\square$

The following lemma contains the essential step to the structure theorem.

Lemma 2.2. *Suppose that* $0 \neq f \in \text{Reg}(A, M)$, *and* M *is indecomposable. Then* $\text{End}(M_R)$ *is a division ring and* $A = \text{Ker}(f) \oplus M'$ *where* $M' \cong M$. *Suppose further that* $\text{Hom}_R(\text{Ker}(f), M) \neq 0$. *Then there is a decomposition* $\text{Ker}(f) = K \oplus M''$ *with* $M'' \cong M$.

Proof. By Theorem 1.1 it follows that $\text{End}(M_R)$ is a division ring. In particular f itself is regular and therefore we have a decomposition $A = \text{Ker}(f) \oplus M'$ where $M' \cong \text{Im}(f) \subseteq^{\oplus} M$. As $f \neq 0$ and M is indecomposable, we have $\text{Im}(f) = M$. Suppose now that $\text{Hom}_R(\text{Ker}(f), M) \neq 0$. Then also $\text{Hom}_R(\text{Ker}(f), M') \neq 0$. Let $0 \neq \varphi \in \text{Hom}_R(\text{Ker}(f), M')$, let $\pi : A \to \text{Ker}(f)$ be the projection along M', and let $\iota : M' \to A$ be the insertion. Then $\iota\varphi\pi \in \text{End}(A_R)$ and $f\iota\varphi\pi$ is a nonzero regular map in $\text{Hom}_R(A, M)$. Hence $\text{Im}(f\iota\varphi\pi) = M$ and $\text{Ker}(f\iota\varphi\pi) = \text{Ker}(\varphi) \oplus M' \subseteq^{\oplus} A$. Then $\text{Ker}(\varphi) \subseteq^{\oplus} \text{Ker}(f)$ and $\text{Ker}(f) = \text{Ker}(\varphi) \oplus M''$ where

$$M'' \cong \frac{\text{Ker}(f)}{\text{Ker}(\varphi)} \cong \frac{\text{Ker}(f) \oplus M'}{\text{Ker}(\varphi) \oplus M'} \cong \frac{A}{\text{Ker}(\phi) \oplus M'} = \frac{A}{\text{Ker}(f\iota\varphi\pi)} \cong \text{Im}(f\iota\varphi\pi) \cong M. \qquad\square$$

Corollary 2.3. *Suppose that* $0 \neq f \in \text{Reg}(A, M)$, *and* M *is indecomposable. Then* $\text{End}(M_R)$ *is a division ring and either* $A = K \oplus M_1 \oplus \cdots \oplus M_n$ *where* $M_i \cong M$, *and* $\text{Hom}_R(K, M) = 0$, *or for every* $i \in \mathbb{N}$ *there is a decomposition* $A = K_i \oplus M_1 \oplus \cdots \oplus M_i$ *with* $M_i \cong M$ *and* $K_i = K_{i+1} \oplus M_{i+1}$. *In the latter case* $\sum_{i=1}^{\infty} M_i = \bigoplus_{i=1}^{\infty} M_i$ *and there is a homomorphism* $\rho : A \to \prod_{i=1}^{\infty} M_i$ *such that* $\bigoplus_{i=1}^{\infty} M_i \subseteq \rho(A)$ *and* $\text{Ker}(\rho) = \bigcap_{i=1}^{\infty} K_i$.

Proof. The decompositions are obtained by repeated applications of Lemma 2.2 as follows. Since f is regular we have $A = K_1 \oplus M_1$ where $K_1 = \text{Ker}(f)$ and $M_1 \cong M$. Suppose that $\text{Hom}(K_1, M) \neq 0$. Then by Lemma 2.2 we have $A = K_2 \oplus M_2 \oplus M_1$, where $K_1 = K_2 \oplus M_2$ and $M_2 \cong M$. Suppose that inductively, $A = K_n \oplus M_n \oplus \cdots \oplus M_2 \oplus M_1$ where $K_n \oplus M_n = K_{n-1} \subseteq \text{Ker}(f)$, $M_i \cong M$, and f is injective on M_1. If $\text{Hom}_R(K_n, M) = 0$, then the theorem is proved. So suppose that $\text{Hom}_R(K_n, M) \neq 0$. Then $\text{Hom}_R(K_n, M_1) \neq 0$. Let $0 \neq \phi \in \text{Hom}_R(K_n, M_1)$. Then $\iota\phi\pi \in \text{End}(A_R)$ where $\pi : A \to M_1$ is the projection along $K_1 = \text{Ker}(f)$ and $\iota : M_1 \to A$ is the insertion. Then $f' := f\iota\phi\pi : A \to M$ is regular and $f'(M_1) = f\iota\phi\pi(M_1) = f\phi(M_1) \neq 0$ because $0 \neq \phi(M_1) \subseteq M_1$ and f is injective on M_1. It follows that $\text{Im}(f') = M$ and $\text{Ker}(f') = M_{n-1} \oplus \cdots \oplus M_2 \oplus \text{Ker}(\phi) \subseteq^{\oplus} A$

and hence $K_{n+1} := \text{Ker}(\phi) \subseteq^{\oplus} K_n$, so $K_n = K_{n+1} \oplus M_{n+1}$ for some submodule M_{n+1} of A. Thus we have $A = K_{n+1} \oplus M_{n+1} \oplus \cdots \oplus M_2 \oplus M_1$ and finally $M_{n+1} \cong \frac{A}{K_{n+1} \oplus M_n \oplus \cdots \oplus M_1} = \frac{A}{\text{Ker}(f')} \cong \text{Im}(f') = M$.

This settles the first case in which the process terminates. In the second case it is easy to see that the sum $\sum_{i=1}^{\infty} M_i$ is direct. The map ρ is obtained as follows. Let $a \in A$ and write $a = k_i + m_1 + \cdots + m_k$ according to the decomposition

$$A = K_i \oplus M_1 \oplus \cdots \oplus M_k$$

the components $m_1, \ldots, m_i$, $i \leq k$ do not depend on the choice of k. This is because $K_i = K_{i+1} \oplus M_{i+1}$. We therefore obtain a well-defined map $\rho(a) = (m_1, \ldots, m_i, \ldots)$ that is obviously homomorphic. Clearly, $\text{Ker}(\rho) = \bigcap_{i=1}^{\infty} K_i$ and $\bigoplus_{i=1}^{\infty} M_i \subseteq \text{Im}(\rho)$. $\qquad\square$

The converse of Corollary 2.3 need not hold.

Example 2.4. There exist abelian groups $A = A_1 \oplus A_2$ and M such that $M \cong A_2$ are indecomposable groups whose endomorphism rings are division rings, and $\text{Hom}(A_1, A_2) = 0$ but $\text{Reg}(A, M) = 0$.

Proof. Let $A_1 \cong \mathbb{Z}(p^{\infty})$, $A_2 \cong \mathbb{Q}$, $M \cong \mathbb{Q}$, and set $A := A_1 \oplus A_2$. Then $M \cong A_2 \cong \mathbb{Q}$ are indecomposable groups whose endomorphism ring is a division ring, namely $\mathbb{Q}$. Also $\text{Hom}(A_1, A_2) = \text{Hom}(\mathbb{Z}(p^{\infty}), \mathbb{Q}) = 0$ but $\text{Hom}(A_2, A_1) = \text{Hom}(\mathbb{Q}, \mathbb{Z}(p^{\infty}))$ contains an epimorphism ϕ. For the purpose of showing that $\text{Reg}(A, M) = 0$, suppose that $\xi = [\xi_{11}, \xi_{21}] \in \text{Reg}(A, M)$, $\xi_{11} \in \text{Hom}(A_1, M)$, $\xi_{21} \in \text{Hom}(A_2, M) \cong \text{Hom}(\mathbb{Q}, \mathbb{Q})$ (where $\xi_{11} = 0$ actually). By Corollary II.6.15 we obtain that $\text{Ker}(\phi\xi_{21}) \subseteq^{\oplus} M \cong \mathbb{Q}$. Suppose that $\xi_{21} \neq 0$. Then it is an isomorphism and $\phi\xi_{21} \neq 0$. As $M \cong \mathbb{Q}$ is indecomposable, the kernel $\text{Ker}(\phi\xi_{21})$ is either 0 or M, neither of which is possible. We conclude that $\xi = 0$. $\qquad\square$

By way of converse to Corollary 2.3 we are left with the hope that $\text{Reg}(A, M) \neq 0$ if $A = M_1 \oplus \cdots \oplus M_n$, $M_i \cong M$, and $\text{End}(M_R)$ is a division ring. The answer is yes.

Theorem 2.5. *Let M_0 be an R-module whose endomorphism ring $D := \text{End}_R(M_0)$ is a division ring. Let $A = A_1 \oplus \cdots \oplus A_m$ and $M = M_1 \oplus \cdots \oplus M_n$ with $A_i \cong M_j \cong M_0$. Then $\text{Reg}(A, M) = \text{Hom}_R(A, M)$.*

Proof. We observe that $D \cong \text{Hom}_R(M, M) \cong \text{Hom}_R(A_i, M_j) \cong \text{End}_R(A_i) \cong \text{End}_R(M_j)$ and all these groups are regular as in any of these all nonzero maps are isomorphisms. Now $\text{Hom}_R(A, M)$, $\text{End}(A_R)$ and $\text{End}(M_R)$ may be considered groups and rings of matrices with entries $\alpha_{si} \in \text{Hom}(A_s, A_i)$, $\xi_{ij} \in \text{Hom}_R(A_i, M_j)$, and $\mu_{jt} \in \text{Hom}(M_j, M_t)$. Every map $\mu_{jt}\xi_{ij}\alpha_{si} \in \text{Hom}_R(A_s, M_t)$ is regular hence every map in $\text{Hom}_R(A, M)$ is regular by Theorem II.6.17. $\qquad\square$

We now consider the case that A is indecomposable and there is a certain abundance of regular maps.

Lemma 2.6. *Suppose that* $0 \neq f \in \operatorname{Hom}(A, M)$, *for every* $\mu \in \operatorname{End}(M_R)$ *the composite* $\mu f \in \operatorname{Hom}_R(A, M)$ *is regular, and* A *is indecomposable. Then* $\operatorname{End}(A_R)$ *is a division ring and* $M = A' \oplus L$ *with* $A' \cong A$. *Suppose further that* $\operatorname{Hom}_R(A, L) \neq 0$. *Then there is a decomposition* $L = A'' \oplus L'$ *with* $A'' \cong A$.

Proof. By Theorem 1.1 we have that $\operatorname{End}(A_R)$ is a division ring. In particular f itself is regular and therefore we have $A \neq \operatorname{Ker}(f) \subseteq^{\oplus} A$ and since A is indecomposable $\operatorname{Ker}(f) = 0$. Also $A \cong \operatorname{Im}(f) \subseteq^{\oplus} M$, hence $M = A' \oplus L$ for some L and $A' \cong A$. As $\operatorname{Hom}_R(A, L) \neq 0$, also $\operatorname{Hom}_R(A', L) \neq 0$. Let $0 \neq \psi \in \operatorname{Hom}_R(A', L)$ and let $\pi : M \to A'$ be the projection along L and let $\iota : L \to M$ be the insertion. Then $\iota\psi\pi \in \operatorname{End}(M_R)$ and $\iota\psi\pi f$ is a regular map in $\operatorname{Hom}_R(A, M)$. Note that $\iota\psi\pi f A = \iota\psi\pi A' = \iota\psi A' = \psi A' \neq 0$. Hence $\iota\psi\pi f \neq 0$ and therefore injective because A is indecomposable. We have further that $A'' := (\iota\psi\pi f)(A) \subseteq^{\oplus} M$ and as $A'' \subseteq L$, hence we get $L = A'' \oplus L'$ for some L' with $A'' \cong A$. $\qquad \square$

Corollary 2.7. *Suppose that* $0 \neq f \in \operatorname{Reg}(A, M)$ *is regular, and* A *is indecomposable. Then* $\operatorname{End}(A_R)$ *is a division ring, and either* $M = A_1 \oplus \cdots \oplus A_n \oplus L$ *where* $A_i \cong A$ *and* $\operatorname{Hom}_R(A, L) = 0$, *or for every* $i \in \mathbb{N}$ *there exist decompositions* $M = A_1 \oplus \cdots \oplus A_i \oplus L_i$ *where* $A_i \cong A$ *and* $L_i = A_{i+1} \oplus L_{i+1}$. *In the latter case* $\sum_{i=1}^{\infty} A_i = \bigoplus_{i=1}^{\infty} A_i$ *and there is a homomorphism* $\rho : M \to \prod_{i=1}^{\infty} A_i$ *such that* $\bigoplus_{i=1}^{\infty} A_i \subseteq \rho(M)$ *and* $\operatorname{Ker}(\rho) = \bigcap_{i=1}^{\infty} L_i$.

Proof. Lemma 2.6 establishes that $\operatorname{End}(A_R)$ is a division ring, and $M = A_1 \oplus L_1$ for some $A_1 \cong A$ and some L_1. If $\operatorname{Hom}_R(A, L_1) = 0$, then we have the claimed decomposition of M. If $\operatorname{Hom}_R(A, L_1) \neq 0$, then Lemma 2.6 tells us that $L_1 = A_2 \oplus L_2$ for some $A_2 \cong A$ and some L_2. We assume by induction that $M = A_1 \oplus \cdots \oplus A_i \oplus L_i$ with $A_i \cong A$. If $\operatorname{Hom}_R(A, L_i) \neq 0$, then Lemma 2.6 tells us that $L_i = A_{i+1} \oplus L_{i+1}$ for some $A_{i+1} \cong A$ and some L_{i+1}. This shows that we have the two claimed possibilities. If the process does not terminate, then we obtain an infinite set $A_1, A_2, \dots$. In this case $\sum_{i=1}^{\infty} A_i = \bigoplus_{i=1}^{\infty} A_i$ because uniqueness of the representation $x = \sum_{i=1}^{\infty} x_i \in \sum_{i=1}^{\infty} A_i$ has finite character.

Suppose that $M = A_1 \oplus \cdots \oplus A_i \oplus L_i$ where $A_i \cong A$ and $L_i = A_{i+1} \oplus L_{i+1}$ for all $i \in \mathbb{N}$. The map ρ is obtained as follows. Let $m \in M$ and write $m = a_1 + \cdots + a_k + \ell_k$ according to the decomposition

$$M = A_1 \oplus \cdots \oplus A_k \oplus L_k,$$

then the components $m_1, \dots, m_i$, $i \leq k$, do not depend on the choice of k. This is because $L_i = A_{i+1} \oplus L_{i+1}$. We therefore obtain a well-defined map $\rho(m) = (a_1, \dots, a_i, \dots)$ that is obviously homomorphic. Clearly, $\operatorname{Ker}(\rho) = \bigcap_{i=1}^{\infty} L_i$ and $\bigoplus_{i=1}^{\infty} A_i \subseteq \operatorname{Im}(\rho)$. $\qquad \square$

The converse of Corollary 2.7 need not hold.

Example 2.8. There exists an indecomposable abelian group A whose endomorphism ring is a division ring and there exists a group $M = A' \oplus L$ such that $A' \cong A$ and $\operatorname{Hom}(A, L) = 0$, yet $\operatorname{Reg}(A, M) = 0$.

Proof. Let $A \cong \mathbb{Q}$, $A' \cong \mathbb{Q}$, $L \cong \mathbb{Z}$, and $M = A' \oplus L$. Then $\text{End}(A) \cong \mathbb{Q}$ is a division ring and $\text{Hom}(A', L) = 0$. Suppose that $\xi = \begin{bmatrix} \xi_{11} \\ \xi_{12} \end{bmatrix} \in \text{Reg}(A, M)$, $\xi_{11} \in$ $\text{Hom}(A, A')$, $\xi_{12} \in \text{Hom}(A, L)$ (actually $\xi_{12} = 0$). By way of contradiction assume that $\xi_{11} \in \text{Hom}(A, A')$ is not zero. Then ξ_{11} is invertible. Let $0 \neq \phi \in \text{Hom}(L, A')$, such maps being plentiful as L is free. Then $0 \neq \xi_{11}\phi \in \text{Hom}(L, A')$ is regular, so $\mathbb{Z} \cong L \cong \xi_{11}\phi L \subseteq^{\oplus} A' \cong \mathbb{Q}$ which is a contradiction. $\qquad \square$

In the category of torsion-free abelian groups of finite rank every group is the direct sum of finitely many indecomposable groups, more generally, Noetherian modules are the direct sum of finitely many indecomposable submodules. This being fairly common, it is of interest to consider the case when both A and M are finite direct sums of indecomposable submodules.

We denote by $\text{tp}\, A$ the isomorphism class of the R-module A and call it the **type** of A. Thus $A_1, A_2 \in \text{tp}\, A$ means that $A_1 \cong A_2 \cong A$. Let $\mathbb{T}$ denote the class of all types of R-modules. Let $A = \bigoplus_{i \in I} A_i$ be a direct decomposition of A. Let $\text{T}_{\text{cr}}(A)$ be the set of types of the direct summands of A. For each type $\tau \in \text{T}_{\text{cr}}(A)$ we can then collect the direct summands into τ-**homogeneous components** $A_\tau := \bigoplus \{A_i \mid A_i \in \tau\}$ and obtain the **homogeneous decomposition** $A = \bigoplus_{\rho \in \text{T}_{\text{cr}}(A)} A_\rho$.

Remark. Note that $\text{T}_{\text{cr}}(A)$ depends on the given decomposition of A and not just on A. If Mod-R happens to be a Remak-Krull-Schmidt category, i.e., a category with unique direct decompositions with indecomposable direct summands, then $\text{T}_{\text{cr}}(A)$ is an invariant of A for each R-module. More generally, $\text{T}_{\text{cr}}(A)$ is an invariant of A if the individual module A has a unique decomposition as a sum of indecomposable submodules.

We first deal with the case when both A and M are "homogeneous".

Lemma 2.9. *Suppose A_0 and M_0 are directly indecomposable modules. Let $A = A_1 \oplus \cdots \oplus A_m$, where $A_i \cong A_0$ for all i, and let $M = M_1 \oplus \cdots \oplus M_n$, where $M_j \cong M_0$ for all j. Suppose that $\text{Reg}(A, M) \neq 0$. Then $A_0 \cong M_0$, $\text{End}_R(A_0) \cong \text{End}_R(M_0)$ are division rings and $\text{Reg}(A, M) = \text{Hom}(A, M)$.*

Proof. Let

$$0 \neq \xi = \begin{bmatrix} \xi_{11} & \cdots & \xi_{m1} \\ \vdots & \vdots & \vdots \\ \xi_{1n} & \cdots & \xi_{mn} \end{bmatrix} \in \text{Reg}(A, M).$$

As $\xi \neq 0$, there is an entry $\xi_{ij} \neq 0$, and $\mu_{jt}\xi_{ij}\alpha_{si} \in \text{Reg}(A_s, M_t)$ for every $\mu_{jt} \in \text{Hom}_R(M_j, M_t)$ and every $\alpha_{si} \in \text{Hom}_R(A_s, A_i)$ (Theorem II.6.17). In particular, for $t = j$, $s = i$, $\mu_{jj} = 1_{M_j}$, $\alpha_{ii} = 1_{A_i}$ we get that $0 \neq \xi_{ij} \in \text{Reg}(A_i, M_j)$. It follows from Proposition 2.1 that $A_0 \cong A_i \cong M_j \cong M_0$ and $\text{End}(A_0) \cong \text{End}(M_0)$ are division rings. By Corollary II.6.19, $\text{Reg}(A, M) = \text{Hom}_R(A, M)$. $\qquad \square$

We recall that an isomorphism class of modules is called a type and denoted by $\sigma, \tau, \rho, \ldots$. We will use $\operatorname{End}\tau$ to denote the isomorphism class of the endomorphism ring of a group in τ.

Theorem 2.10. *Suppose that $A = A_1 \oplus \cdots \oplus A_m$ and $M = M_1 \oplus \cdots \oplus M_n$ are decompositions with indecomposable direct summands. Let $A = \bigoplus_{\rho \in \mathrm{T_{cr}}(A)} A_\rho$ and $M = \bigoplus_{\rho \in \mathrm{T_{cr}}(M)} M_\rho$ be the associated homogeneous decompositions. Suppose that $0 \neq \xi = [\xi_{\sigma\tau}] \in \operatorname{Reg}(A, M)$ where $\xi_{\sigma\tau} \in \operatorname{Hom}_R(A_\sigma, M_\tau)$. Suppose that $\xi_{\sigma\tau} \neq 0$. Then the following statements hold.*

1) $\sigma = \tau$,

2) $\operatorname{End}(\sigma) = \operatorname{End}(\tau)$ *are division rings,*

3) $\operatorname{Hom}_R(A_{\sigma'}, A_\sigma) = 0$ *whenever $\sigma' \neq \sigma$,*

4) $\operatorname{Hom}_R(M_\tau, M_{\tau'}) = 0$ *whenever $\tau' \neq \tau$.*

Proof. 1), 2) By Theorem II.6.17 we have that $\mu_{\tau t} \xi_{\sigma\tau} \alpha_{s\sigma} \in \operatorname{Reg}(A_s, M_t)$ for all t, s, all $\mu_{\tau t} \in \operatorname{Hom}_R(M_\tau, M_t)$, and all $\alpha_{s\sigma} \in \operatorname{Hom}_R(A_s, A_\sigma)$. In particular, for $t = \tau$, $s = \sigma$, $\mu_{\tau\tau} = 1$, $\alpha_{\sigma\sigma} = 1$ we find that $0 \neq \xi_{\sigma\tau} \in \operatorname{Reg}(A_\sigma, M_\tau)$. By Lemma 2.9 we conclude that $\sigma = \tau$, $\operatorname{End}(\sigma) = \operatorname{End}(\tau)$ is a division ring and $\operatorname{Reg}(A_\sigma, M_\tau) = \operatorname{Hom}_R(A_\sigma, M_\tau)$.

3) We can write $\xi = [\xi_{ij}]$ where $\xi_{ij} \in \operatorname{Hom}_R(A_i, M_j)$. As $\xi_{\sigma\tau} = \xi_{\sigma\sigma} \neq 0$ there exist i and j such that $\xi_{ij} \neq 0$ and $A_i, M_j \in \sigma$. Also, by Theorem II.6.17, $\xi_{ij} \in \operatorname{Reg}(A_i, M_j)$ and this implies that ξ_{ij} is bijective (Proposition 2.1). Now suppose that $0 \neq \alpha_{si} \in \operatorname{Hom}_R(A_s, A_i)$. Then $\xi_{ij}\alpha_{si} \in \operatorname{Reg}(A_s, M_j)$ and $\xi_{ij}\alpha_{si} \neq 0$. It follows again by Proposition 2.1 that $A_s \cong M_j \cong A_i$. This means that $\operatorname{Hom}_R(A'_\sigma, A_\sigma) = 0$ whenever $\sigma' \neq \sigma$.

4) Again we use $\xi = [\xi_{ij}]$ where $\xi_{ij} \in \operatorname{Hom}_R(A_i, M_j)$ and assume that $\xi_{ij} \neq 0$ where $A_i, M_j \in \tau$, and hence ξ_{ij} is bijective. Now suppose that $0 \neq \mu_{jt} \in \operatorname{Hom}_R(M_j, M_t)$. Then $\mu_{jt}\xi_{ij} \in \operatorname{Reg}(A_i, M_t)$ and $\mu_{jt}\xi_{ij} \neq 0$. It follows again by Proposition 2.1 that $M_t \cong A_i \cong M_j$. This means that $\operatorname{Hom}_R(M_\tau, M_{\tau'}) = 0$ whenever $\tau' \neq \tau$. $\qquad\square$

We have the following converse.

Theorem 2.11. *Suppose that $A = A_1 \oplus \cdots \oplus A_m$ and $M = M_1 \oplus \cdots \oplus M_n$ are decompositions with indecomposable direct summands. Let $A = \bigoplus_{\rho \in \mathrm{T_{cr}}(A)} A_\rho$ and $M = \bigoplus_{\rho \in \mathrm{T_{cr}}(M)} M_\rho$ be the associated homogeneous decompositions. Let $\Sigma \subseteq \mathrm{T_{cr}}(A) \cap \mathrm{T_{cr}}(M)$. Suppose that $\xi = [\xi_{\sigma\tau}] \in \operatorname{Hom}_R(A, M)$ is such that*

1) $\xi_{\sigma\tau} = 0$ *unless $\sigma = \tau$,*

2) $\xi_{\sigma\sigma} \in \operatorname{Reg}(A_\sigma, M_\sigma)$ *for $\sigma \in \Sigma$,*

3) $\xi_{\sigma\sigma} = 0$ *for $\sigma \notin \Sigma$,*

4) *whenever $\sigma' \neq \sigma$, then $\operatorname{Hom}_R(A_{\sigma'}, A_\sigma) = 0$,*

5) *whenever $\tau' \neq \tau$, then $\mathrm{Hom}_R(M_\tau, M_{\tau'}) = 0$.*

Then $\xi \in \mathrm{Reg}(A, M)$.

Proof. We need to show that $\mu_{\sigma\sigma''}\xi_{\sigma\sigma}\alpha_{\sigma'\sigma} \in \mathrm{Reg}(A_{\sigma'}, M_{\sigma''})$ for all $\sigma \in \Sigma$, $\sigma' \in \mathrm{T}_{\mathrm{cr}}(A)$ and $\sigma'' \in \mathrm{T}_{\mathrm{cr}}(M)$. Let $\sigma \in \Sigma$ be given. If $\sigma' \neq \sigma$ or $\sigma'' \neq \sigma$, then $\mu_{\sigma\sigma''}\xi_{\sigma\sigma}\alpha_{\sigma'\sigma} = 0 \in \mathrm{Reg}(A_{\sigma'}, M_{\sigma''})$. So suppose that $\sigma' = \sigma'' = \sigma$. Then $\mu_{\sigma\sigma}\xi_{\sigma\sigma}\alpha_{\sigma\sigma} \in \mathrm{Reg}(A_\sigma, M_\sigma)$ because $\xi_{\sigma\sigma} \in \mathrm{Reg}(A_\sigma, M_\sigma)$ by hypothesis. $\square$

Chapter IV

Regularity in Modules

1 Fundamental Results

A very interesting special case of our general concept brings us to the regularity of modules. In $\mathrm{Hom}_R(A, M)$ we take $A = R_R$. We then have $T := \mathrm{End}(R_R) = R$ where R acts by left multiplication on R and we obtain the S-R-bimodule $\mathrm{Hom}_R(R, M)$ where $S := \mathrm{End}(M_R)$. Of course, M also is an S-R-bimodule. The first basic observation that allows us to transfer our previous more general results to the module M is the routine fact that

$$\rho : \mathrm{Hom}_R(R, M) \ni f \mapsto f(1) \in M \tag{1}$$

is a bimodule isomorphism.

We now come to the definition of regularity for modules. The map $f \in \mathrm{Hom}_R(R, M)$ is regular if there exists $\varphi \in \mathrm{Hom}_R(M, R)$ such that

$$f\varphi f = f. \tag{2}$$

Applying ρ to $f\varphi f = f$, i.e., evaluating at $1 \in R$, we obtain

$$m := f(1) = (f\varphi f)(1) = f(\varphi(m)) = f(1)\varphi(m) = m\varphi(m),$$

so $m = m\varphi(m)$. Hence the previous definition reappears in the module setting as follows.

Definition 1.1. An element $m \in M$ is **regular** with **quasi-inverse** $\varphi \in \mathrm{Hom}_R(M, R)$ if

$$m = m\varphi(m). \tag{3}$$

This means that $f \in \mathrm{Hom}_R(R, M)$ *is regular in* $\mathrm{Hom}_R(R, M)$ *if and only if* $f(1) \in M$ *is regular in* M.

Several authors have studied regular modules ([37], [10], [35]).

Regular modules are special as the following example shows.

Theorem 1.2. *Let $M \neq 0$ be a regular module over an integral domain R. Then R is a field. A vector space over a division ring is regular.*

Proof. Let R be an integral domain and M a nonzero regular R-module. Then for every $0 \neq a \in M$ there is $\varphi \in \operatorname{Hom}_R(M, R)$ such that $a\varphi(a) = a$. Suppose that a is a torsion element, i.e., there is $0 \neq r \in R$ such that $ar = 0$. Then $0 = \varphi(ar) = \varphi(a)r$ and since R is an integral domain, $\varphi(a) = 0$ and therefore $a = a\varphi(a) = 0$. So M contains no nonzero torsion elements, i.e., M is a torsion-free module. It now follows from $a\varphi(a) = a \neq 0$ that $a(\varphi(a) - 1) = 0$ and therefore $\varphi(a) = 1$. Hence $\varphi : M \to R$ is surjective and since R is projective it follows that $M = aR \oplus K$ for every nonzero $a \in M$. Let $0 \neq s \in R$. Then, as for a, we have $M = asR \oplus L$ for some L. As $asR \subseteq aR$ we can intersect $M = asR \oplus L$ with aR to get $aR = asR \oplus (K \cap aR)$. Then $K \cap aR \cong aR/asR$ and $(K \cap aR)s \cong (aR/asR)\, s = 0$ and since M is torsion-free, we conclude that $K \cap aR = 0$ and $aR = asR$. Hence there exists $t \in R$ such that $a = ast$, so $1 = st$, showing that s is invertible. Thus R is a field.

It is easy to see that vector spaces are regular modules. $\square$

By means of the isomorphism (1) we can specialize our general results to immediately obtain numerous results on regularity in modules.

Since we will have to work with $f \in \operatorname{Hom}_R(R, M)$ we mention first some simple properties of such an f. The image of f is

$$\operatorname{Im}(f) = f(R) = f(1)R.$$

Since $f(1) \in M$, we set $m := f(1)$. Then

$$\operatorname{Im}(f) = f(R) = mR.$$

The kernel of f is

$$\operatorname{Ker}(f) := \{r \in R \mid f(r) = mr = 0\} = \operatorname{Ann}_R(m),$$

the annihilator of m in R.

For $f \in \operatorname{Hom}_R(R, M)$ and $\varphi \in \operatorname{Hom}_R(M, R)$ we have $f\varphi \in S = \operatorname{End}(M_R)$. With $m := f(1)$ we have for all $x \in M$ that

$$f\varphi(x) = f(1 \cdot \varphi(x)) = f(1)\varphi(x) = m\varphi(x).$$

This motivates the following definition of the product of a module element $m \in M$ and a map $\varphi \in \operatorname{Hom}_R(M, R)$,

$$m\varphi : M \ni x \mapsto m\varphi(x) \in M.$$

It is easily checked that $m\varphi \in S = \operatorname{End}(M_R)$.

In general, if $fgf = f$, then we have idempotents $e := fg \in S$ and $d := gf \in T$. Hence for $f \in \operatorname{Hom}_R(R, M)$ with $f(1) = m$ and $\varphi \in \operatorname{Hom}_R(M, R)$ we have the idempotents

$$e := m\varphi \in S \quad \text{and} \quad d := \varphi(m) \in R.$$

Corresponding to Theorem II.1.2 we now have the following decomposition theorem for regular elements in a module.

Theorem 1.3. *If $m \in M$ is regular with quasi-inverse $\varphi \in \mathrm{Hom}_R(M, R)$, then*

$$R = \mathrm{Ann}_R(m) \oplus \varphi(m)R, \quad \varphi(m)R \cong mR = m(\mathrm{Im}(\varphi)), \qquad (4)$$

and

$$M = mR \oplus \{x \in M \mid \varphi(x) \in \mathrm{Ann}_R(m)\}.$$

Furthermore, the mapping

$$m_0 : \varphi(m)R \ni \varphi(m)r \mapsto m\varphi(m)r = mr \in mR$$

is an isomorphism with inverse isomorphism

$$\varphi_0 : mR \ni mr \mapsto \varphi(m)r \in \varphi(m)R.$$

Proof. This is just Theorem II.1.2 with $\mathrm{Ker}(f) = \mathrm{Ann}_R(m)$, $\mathrm{Im}(f) = mR$, $\mathrm{Im}(\varphi f) = \varphi f(R) = \varphi(mR) = \varphi(m)R$, and $\mathrm{Im}(f\varphi) = f\varphi(M) = m\,\mathrm{Im}(\varphi)$. $\qquad\square$

The following theorem is essentially a specialization of Corollary II.1.3.

Theorem 1.4 (Characterization of Regularity). *Let $m \in M$. Then the following statements are equivalent.*

1) *m is regular.*

2) *$\mathrm{Ann}_R(m) \subseteq^\oplus R$ and $mR \subseteq^\oplus M$.*

3) *mR is an R-projective direct summand of M.*

Proof. The equivalence of 1) and 2) follows from Corollary II.1.3, and 3) follows from 2) because mR is isomorphic to the complementary direct summand of $\mathrm{Ann}_R(m)$ in R. To finish, assume that mR is projective. We have the epimorphism $R \twoheadrightarrow mR$ and since mR is projective, the kernel $\mathrm{Ann}_R(m)$ is a direct summand of R. Hence 3) implies 2). $\qquad\square$

Corresponding to Corollary II.1.5 we have:

Corollary 1.5. *The module M contains a regular element $\neq 0$ if and only if there exist nonzero direct summands of R and of M that are isomorphic.*

From Corollary II.1.4 we obtain the following cases of regular modules.

Corollary 1.6. *In the following three cases all elements of M are regular.*

1) *R and M are both semisimple.*

2) *R is semisimple and injective.*

3) *M is semisimple and projective.*

Further examples of regular modules are obtained ([37, page 343]) as right ideals in regular rings. Rings of linear transformations are regular rings and the ring of linear transformations of a countable-dimensional vector space contains nonprojective right ideals.

Now we will show that in every nonzero module there exists a largest regular submodule.

Definition 1.7. Let M_R be a given module. Let

$$\text{Reg}(M_R) := \{m \in M \mid SmR \text{ is regular}\}.$$

Theorem 1.8. $\text{Reg}(M_R)$ *is the largest regular S-R-submodule of M.*

Proof. Theorem II.4.3 and $\rho(\text{Reg}(R, M)) = \text{Reg}(M_R)$. □

Question 1.9. Describe all modules M_R with $\text{Reg}(M_R) = M$ and $\text{Reg}(M_R) = 0$. Zelmanowitz's paper is all about the case $\text{Reg}(M_R) = M$. The case $\text{Reg}(M_R) = 0$ may be unmanageable. What are the consequences of the assumption that $\text{Reg}(M_R) \neq 0$?

2 Quasi-Inverses

2.1 Basic Properties

In considering regularity in modules there is now a distinct asymmetry between M and $\text{Hom}_R(M, R)$. To emphasize this asymmetry we set $M^* := \text{Hom}_R(M, R)$ and use Greek letters $\varphi, \psi, \eta, \ldots$ for its elements. Note that M is an S-R-bimodule while M^* is an R-S-bimodule where $S = \text{End}(M_R)$ and $\text{End}(R_R) = R$ (by left multiplication).

We observe that in addition to regular elements $m \in M$ we also have regular maps $\varphi \in M^*$.

1) $m \in M$ is regular with quasi-inverse $\varphi \in M^*$ if and only if $m = m\,\varphi(m)$.

2) $\varphi \in M^*$ is regular with quasi-inverse m if and only if $\varphi = \varphi(m)\varphi$.

In general, it follows from $fgf = f$ that

$$f(gfg)f = f, \qquad (gfg)f(gfg) = gfg,$$

the interesting point being that gfg is a quasi-inverse of f that is again regular and has quasi-inverse f.

Suppose now that $f \in \text{Hom}_R(R, M)$, $f(1) = m$ and $\varphi \in \text{Hom}_R(M, R)$ is a quasi-inverse of f or, equivalently, of m. Then $(\varphi f \varphi)(x) = \varphi(m\varphi(x)) = \varphi(m)\varphi(x)$, i.e., $\varphi f \varphi = \varphi(m)\varphi$. Hence $\varphi(m)\varphi$ must be a quasi-inverse of m that is itself regular with quasi-inverse m. We will verify this directly. Suppose that $m\varphi(m) = m$. Then applying φ to $m\varphi(m) = m$ we have that $\varphi(m)^2 = \varphi(m)$ and now

$$m(\varphi(m)\varphi)(m) = m\varphi(m)^2 = m\varphi(m) = m,$$

and

$$(\varphi(m)\varphi)(m)(\varphi(m)\varphi) = \varphi(m)^3\varphi = \varphi(m)\varphi.$$

This shows that m and $\varphi(m)\varphi$ are both regular and mutual quasi-inverses.

The case of regular objects that are mutal quasi-inverses is particularly nice.

Proposition 2.1. *The following statements hold.*

1) *Let $m \in M$ be regular. Then there exists a quasi-inverse φ of m that is also regular and has quasi-inverse m, i.e., in addition to $m\varphi(m) = m$ we have that $\varphi(m)\varphi = \varphi$.*

2) *Suppose that $m \in M$ and $\varphi \in M^* = \operatorname{Hom}_R(M, R)$ are such that $m\varphi(m) = m$ and $\varphi(m)\varphi = \varphi$. Then*

$$R = \operatorname{Ann}_R(m) \oplus \operatorname{Im}(\varphi), \quad and \quad M = mR \oplus \operatorname{Ker}(\varphi).$$

Proof. Proposition II.1.6. □

2.2 Partially Invertible Objects are Quasi-Inverses

If $m \in M$ is regular and $m\varphi(m) = m$ for $\varphi \in M^* = \operatorname{Hom}_R(M, R)$, then we have the idempotents

$$e := m\varphi \in S, \quad d := \varphi(m) \in R,$$

and m and φ are "factors" of $d = \varphi m = \varphi(m)$ and of $e = m\varphi$. Conversely, a "factor" of an idempotent is the quasi-inverse of some regular map as we will see next. The result that we specialize here is Theorem II.2.1.

Due to the asymmetry inherent in our present set-up, Theorem II.2.1 splits into two parts. We first deal with the case that $\varphi \in M^*$ is a factor of an idempotent.

Lemma 2.2. *The following statements are equivalent for $\varphi \in \operatorname{Hom}_R(M, R)$.*

1) *There exists $m \in M$ such that $m\varphi = (m\varphi)^2 \neq 0$.*

2) *There exists $n \in M$ such that $\varphi(n) = \varphi(n)^2 \neq 0$.*

3) *There exists $m \in M$ such that $m\varphi(m) = m \neq 0$.*

4) *There exist $0 \neq M_0 \subseteq^{\oplus} M$, and $R_0 \subseteq^{\oplus} R$ such that*

$$\varphi_0 : M_0 \ni x \mapsto \varphi(x) \in R_0$$

is an isomorphism.

Next we consider the case when $m \in M$ is a factor of an idempotent.

Lemma 2.3. *The following statements are equivalent for $m \in M$.*

1) *There exists $\psi \in M^*$ such that $m\psi = (m\psi)^2 \neq 0$.*

2) *There exists $\eta \in M^*$ such that $\eta(m) = \eta(m)^2 \neq 0$.*

3) *There exists $\varphi \in M^*$ such that $m\varphi(m) = m \neq 0$.*

4) *There exist $0 \neq M_0 \subseteq^{\oplus} M$, and $R_0 \subseteq^{\oplus} R$ such that*

$$m_0 : R_0 \ni x \mapsto mx \in M_0$$

is an isomorphism.

Definition 2.4.

1) $\varphi \in M^*$ is **partially invertible** if the equivalent conditions of Lemma 2.2 are satisfied.

2) $m \in M$ is **partially invertible** if the equivalent conditions of Lemma 2.3 are satisfied.

3) The **total** of $\varphi = \text{Hom}_R(M, R)$ is defined to be

$$\text{Tot}(M, R) := \{\varphi \in M^* \mid \varphi \text{ is not partially invertible.}\}.$$

4) The **total** of the right R-module M is defined to be

$$\text{Tot}(M_R) := \{m \in M \mid m \text{ is not partially invertible.}\}.$$

For an arbitrary module M, one of the following statements is true ([20, page 28]).

1) $M = \text{Tot}(M)$.

2) $M = U \oplus \bigoplus_{i=1}^{n} m_i R$ with $n \geq 1$, $U \subseteq \text{Tot}(M)$, m_i regular, and $m_i R$ projective for $i = 1, \ldots, n$.

3) M contains a finitely direct summand of the form $\bigoplus_{i \in \mathbb{N}} m_i R$ where the $m_i R$ have the same properties as in 2).

We mention two results on uniqueness of quasi-inverses that are the specializations of Proposition II.2.6 and Corollary II.2.7.

Proposition 2.5. *Suppose that $m \in M$ is regular and $\text{End}(M_R)$ is commutative. Then m has a unique quasi-inverse.*

Proposition 2.6. *Let $m \in M$ be regular with quasi-inverse $\varphi \in \text{Hom}_R(M, R)$ and suppose that R is commutative. Then m is the unique quasi-inverse of $\varphi(m)\varphi$.*

3   Regular Elements Generate Projective Direct Summands

Here again we obtain interesting results on regular module elements by specializing general results.

Theorem 3.1. *Let $0 \neq m \in M$ be regular and set $S = \mathrm{End}(M_R)$. Then:*

1) *Sm is a nonzero S-projective direct summand of $_SM$ that is isomorphic to a cyclic left ideal of S that is a direct summand of S.*

2) *mR is a nonzero R-projective direct summand of M_R that is isomorphic to a cyclic right ideal of R that is a direct summand of R.*

More precisely, let $\varphi \in \mathrm{Hom}_R(M, R)$ be a quasi-inverse of m. Then:

1) *$M = Sm \oplus M(1-\varphi(m))$, $S = Sm\varphi \oplus S(1-m\varphi)$, and $Sm = M\varphi(m) \cong Sm\varphi$.*

2) *$R = \varphi(m)R \oplus (1 - \varphi(m))R$, $M = mR \oplus (1 - m\varphi)M$, and $mR \cong \varphi(m)R$.*

Proof. Apply ρ to the corresponding result on $_S\mathrm{Hom}_R(R, M)$ (Theorem II.3.1). □

Until now we considered M as an S-R-bimodule. We set

$$S := \mathrm{End}(M_R), \quad E := \mathrm{End}(_SM).$$

Then M is also an S-E-bimodule.

Corollary 3.2. *If $0 \neq m \in M$ is regular, then*

1) *$Sm \neq 0$ is an S-projective direct summand of $_SM$.*

2) *$mE \neq 0$ is an E-projective direct summand of M_E.*

Proof. 1) is nothing new. It coincides with Theorem 3.1.1.

2) follows from the more general Corollary II.3.2. □

These results have implications for the structure of $\mathrm{Reg}(M_R)$.

Theorem 3.3. *Let N be a finitely generated S-submodule of $\mathrm{Reg}(M_R)$. Then N is an S-projective direct summand of M. Furthermore, N is the direct sum of finitely many cyclic projective submodules, each of which is isomorphic to a left ideal of S. The same is true for submodules of $\mathrm{Reg}(M_R)$ considered as a right R-module.*

Proof. This is obtained by either applying ρ to Theorem II.4.6 or as a corollary of Theorem 3.1 using Lemma II.4.5. □

Theorem 3.4. *Suppose that $\mathrm{Reg}(M_R)$ contains no infinite direct sums of S-submodules. Then $\mathrm{Reg}(M_R)$ is the direct sum of finitely many simple projective S-modules, $M = \mathrm{Reg}(M_R) \oplus U$ for some S-submodule U and U contains no nonzero regular S-R-submodule. Every cyclic S-submodule of $\mathrm{Reg}(M_R)$ is isomorphic to a left ideal of S that is a direct summand of S.*

An analogous result holds when $\mathrm{Reg}(M_R)_R$ contains no infinite direct sums of R-submodules.

Proof. Corollary II.4.8. □

Question 3.5. When is $\text{Reg}(M_R)$ S-projective, when R-projective?

Question 3.6. Zelmanowitz has the assumption that R is "left self injective". What can be said about $\text{Hom}_R(A, M)$ if A_R is injective or if M_R is projective?

The following result is an immediate corollary of Theorem II.4.11.

Theorem 3.7. *Suppose that* $S := \text{End}(M_R)$ *is left perfect. Then* $\text{Reg}(M_R)$ *is S-projective. If R is right perfect, then $\text{Reg}(M_R)$ is R-projective.*

Theorem 3.8. ([37, Theorem 1.11]) *Suppose that R is finite-dimensional, i.e., R_R contains no infinite direct sums of right ideals. Then every regular R-module is projective.*

Proof. Let M be a regular R-module. By Lemma II.4.5 we have that every finitely generated submodule is a direct summand of M and every finitely or countably generated submodule of M is a direct sum of cyclic regular modules. By Zorn there is a maximal submodule that is the direct sum of cyclic submodules. Let $N = \bigoplus_{a \in A} Rm_a$, $m_a \in M$, be such a maximal submodule. We claim that $N = M$. Let $0 \neq m \in M$. By Theorem 1.3 the cyclic module mR is isomorphic with a right ideal of R, so mR contains no infinite direct sums. By Theorem 3.4 we conclude that $mR = n_1 R \oplus \cdots \oplus n_t R$ with t a natural number and $n_i R$ simple. But Rn_i is simple and $N \cap n_i R \neq 0$ implies that $n_i R \subseteq N$ for every i, hence $m \in mR \subseteq N$, and we have shown that $M \subseteq N$, so $M = N$. $\square$

4 Remarks on the Literature

The paper [37] contains numerous results on regular modules. Many of these results have been generalized in this monograph in two ways: we deal with the bimodule $_S\text{Hom}_R(A, M)_T$ instead of $_SM_R$ and we do not assume that $\text{Hom}_R(A, M)$ is regular but instead make statements about the largest regular submodule $\text{Reg}(A, M)$. This is a generalization because $\text{Hom}_R(A, M)$ is regular if and only if $\text{Hom}_R(A, M) = \text{Reg}(A, M)$. Zelmanowitz extensively discusses endomorphism rings of regular modules which is a topic specific to regular modules and we only cite a few sample results without proof.

The endomorphism ring of a regular module need not be regular but ([37, (3.2)]) it is always semiprime.

Theorem 4.1 ([37, Theorem 3.4]). *If $_RM$ is a regular module, then the center of $\text{End}(_RM)$ is regular.*

Corollary 4.2 ([37, Corollary 3.5]). *If J is a commutative regular left ideal of R, then $\text{End}(_RJ)$ is a commutative regular ring.*

Theorem 4.3 ([37, Theorem 3.8]). *Let R be a commutative ring and suppose that M is an R-module with the property that every cyclic submodule is contained in a projective direct summand of M. If $\text{End}(_RM)$ is a regular ring, then M is a regular module.*

Corollary 4.4 ([37, Corollary 4.2], [35, Theorem 3.6]). *If M is a finitely generated regular module, then $\mathrm{End}(M)$ is a regular ring.*

Theorem 4.5 ([37, Theorem 3.8]). *If $_RM$ is a regular R-module with homogeneous socle, then $\{\alpha \in \mathrm{End}(M) \mid M\alpha \text{ is finite-dimensional}\}$ is an ideal of $\mathrm{End}(M)$ and a simple regular ring.*

Theorem 4.6 ([37, Theorem 3.8]). *Let $_RM$ be a regular module. Then $\mathrm{End}(_RM)$ is semisimple with minimum condition if and only if M is finite-dimensional.*

5 The Transfer Rule

The transfer rule can also be restricted to the module setting.

If $m \in M$ and $\varphi \in \mathrm{Hom}_R(M, R)$ with $m\varphi(m) = m$, then we call (m, φ) a **regular pair**. Similarly, if $\varphi \in \mathrm{Hom}_R(M, R)$ and $k \in M$ with $\varphi(k)\varphi = \varphi$, then (φ, k) is a regular pair. If (m, φ) is a regular pair, then also $(\varphi(m)\varphi, m)$ is a regular pair.

Definition 5.1. If (m, φ) is a regular pair with $m \in M$, then we call

$$\mathrm{trf} : (m, \varphi) \mapsto (\varphi(m)\varphi, m)$$

the **transfer rule** or the **transfer** and we also write $\mathrm{trf}(m, \varphi) = (\varphi(m)\varphi, m)$.

Theorem 5.2. *If trf is applied to all regular pairs (m, φ) with $m \in M$, then the set of first entries in $\mathrm{trf}(m, \varphi) = (\varphi(m)\varphi, m)$ is the set of all regular elements in $\mathrm{Hom}_R(M, R)$.*

Chapter V

Regularity in $\mathrm{Hom}_R(A, M)$ as a One-sided Module

1 Iterated Endomorphism Rings

So far we discussed regularity in $\mathrm{Hom}_R(A, M)$, and then by using $\mathrm{Hom}_R(R, M) \cong M$ we arrived at regularity in modules. We can now look at module structures on $\mathrm{Hom}_R(A, M)$ over certain rings and study regularity of $\mathrm{Hom}_R(A, M)$ as a module over these rings.

Explicitly, we have module structures on $H := \mathrm{Hom}_R(A, M)$ as follows.

- $_S H$ where $S = \mathrm{End}(M_R)$, $\forall s \in S, \forall h \in H, s \cdot h = s \circ h$,

- H_T where $T = \mathrm{End}(A_R)$, $\forall t \in T, \forall h \in H, h \cdot t = h \circ t$,

- $_{S'} H$ where $S' = \mathrm{End}(H_T)$, $\forall s' \in S', \forall h \in H, s' \cdot h = s'(h)$,

- $H_{T'}$ where $T' = \mathrm{End}(_S H)$, $\forall t' \in T', \forall h \in H, h \cdot t' = (h)t'$,

- $_{S''} H$ where $S'' = \mathrm{End}(H_{T'})$, $\forall s'' \in S'', \forall h \in H, s'' \cdot h = s''(h)$,

- $H_{T''}$ where $T'' = \mathrm{End}(_{S'} H)$, $\forall t'' \in T'', \forall h \in H, h \cdot t'' = (h)t''$.

Associated with each of these there is a largest regular bi-submodule and one might think that there is an unending chain of such substructures of H. However, this is not so because the process of forming endomorphism rings comes to a standstill after three steps as we will explain next. This is well-known and can be found in [1, pages 60-61].

Given an R-module M_R we have its endomorphism ring $E_1 := \mathrm{End}(M_R)$, and M becomes a left E_1-module $_{E_1} M$ where left multiplication by $s \in E_1$ is left action of s on M, i.e., for $x \in M$, $s \cdot x = s(x)$. Let $E_2 := \mathrm{End}(_{E_1} M)$ and M become a right E_2-module where again right multiplication by $t \in E_2$ is right action of t. Further we have $E_3 := \mathrm{End} M_{E_2}$. We will show that $E_3 \cong E_1$ and $E_4 := \mathrm{End}(_{E_3} M) \cong E_2$

with identical actions on M so that nothing new is obtained by iterating the process further.

Proposition 1.1. *Let M be a right R-module.*

1) *Set $E_0 := R$ and set $E_1 := \operatorname{End}(M_R) = \operatorname{End}(M_{E_0})$. For $s \in E_1$ and $x \in M$, define $s \cdot x := s(x)$, making M into a bimodule $_{E_1}M_{E_0}$.*

2) *Set $E_2 := \operatorname{End}(_{E_1}M)$. For $t \in E_2$ and $x \in M$, define $x \cdot t := (x)t$, making M into a bimodule $_{E_1}M_{E_2}$.*

3) *$\rho : E_0 \to E_2 : \forall r \in R, \forall x \in M, (x)\rho(r) = x \cdot r$ determines a well-defined ring homomorphism.*

4) *Set $E_3 := \operatorname{End}(M_{E_2})$. For $u \in E_3$ and $x \in M$, define $u \cdot x := u(x)$, making M into a bimodule $_{E_3}M_{E_2}$.*

5) *$\lambda : E_1 \to E_3 : \forall s \in E_1, \forall x \in M, \lambda(s)(x) = s(x)$ determines a well-defined ring isomorphism.*

6) *$\forall x \in M, \forall s \in E_1 : s \cdot x = \lambda(s) \cdot x$, and hence $\operatorname{End}(_{E_3}M) = E_2$.*

7) *Set $E_4 := \operatorname{End}(_{E_3}M)$. For $v \in E_4$ and $x \in M$, define $x \cdot v := (x)v$, making M into a bimodule $_{E_3}M_{E_4}$.*

8) *$\rho : E_2 \to E_4 : \forall t \in E_2, \forall x \in M, (x)\rho(t) = (x)t$ determines a well-defined ring isomorphism.*

9) *$\forall x \in M, \forall t \in E_2 : x \cdot t = x \cdot \rho(t)$, and hence $\operatorname{End}(M_{E_4}) = E_3$.*

Proof. Most statements being evident and well-known, we check only 5) and 6). Clearly λ is a ring homomorphism. Suppose that $\lambda(s) = 0$. Then $\forall x \in M : s(x) = 0$ which means that $s = 0$. Now let $t \in E_3$. We will see that $t : M \to M$ is an R-homomorphism which says that $t \in E_1$ and so $\lambda(t) = t$. We have that

$$\rho : R \to E_2 : \forall r \in R, \forall x \in M, (x)\rho(r) = x \cdot r$$

is a well-defined ring homomorphism. Recall that $t \in E_3$. We compute that $t(xr) = t(x\rho(r)) = t(x)\rho(r) = t(x) \cdot r$ showing that t is indeed an R-homomorphism.

The proofs of 8) and 9) are similar. $\square$

2   Definitions and Characterizations

We already worked with $H := \operatorname{Hom}_R(A, M)$ as an S-T-bimodule where

$$S := \operatorname{End}(M_R), \quad T := \operatorname{End}(A_R)$$

and we have additionally

$$S' := \operatorname{End}(H_T), \quad T' := \operatorname{End}(_S H), \quad S'' := \operatorname{End}(H_{T'}), \quad T'' := \operatorname{End}(_{S'} H).$$

Thus H is an S-T-, an S-T'-, an S'-T-, an S'-T''-, and an S''-T'-module. This is as far as we can go because, by Proposition 1.1, we have

$$S''' := \mathrm{End}(_{S''}H) = S', \quad \text{and} \quad T''' := \mathrm{End}(H_{T''}) = T'.$$

We also have, assuming that A_R is a faithful R-module,

$$R \subseteq T', \quad S \subseteq S'', \quad \text{and} \quad T \subseteq T''.$$

So $H = \mathrm{Hom}_R(A, M)$ is a module in several ways and for each we have regularity concepts and regular substructures.

Let $f \in H := \mathrm{Hom}_R(A, M)$.

- f is **S-regular**, i.e., f is regular as an element of $_S H$, if and only if there exists $\sigma \in \mathrm{Hom}_S(H, S)$ such that $(f\sigma)f = f$.

- f is **T-regular**, i.e., f is regular as an element of H_T, if and only if there exists $\tau \in \mathrm{Hom}_T(H, T)$ such that $f(\tau f) = f$.

- f is **S'-regular**, i.e., f is regular as an element of $_{S'} H$, if and only if there exists $\sigma' \in \mathrm{Hom}_{S'}(H, S')$ such that $(f\sigma')f = f$.

- f is **T'-regular**, i.e., f is regular as an element of $H_{T'}$, if and only if there exists $\tau' \in \mathrm{Hom}_{T'}(H, T')$ such that $f(\tau'f) = f$.

- f is **S''-regular**, i.e., f is regular as an element of $_{S''} H$, if and only if there exists $\sigma'' \in \mathrm{Hom}_{S''}(H, S'')$ such that $(f\sigma'')f = f$.

- f is **T''-regular**, i.e., f is regular as an element of $H_{T''}$, if and only if there exists $\tau'' \in \mathrm{Hom}_{T''}(H, T'')$ such that $f(\tau''f) = f$.

It turns out that the regularity of $f \in \mathrm{Hom}_R(A, M)$ implies all other kinds of regularity.

Lemma 2.1. *Let* $f \in \mathrm{Hom}_R(A, M)$ *be regular and let* $g \in \mathrm{Hom}_R(M, A)$ *be such that* $fgf = f$. *Then the map*

$$\sigma : {}_S\mathrm{Hom}_R(A, M) \ni h \mapsto (h)\sigma := h \circ g \in {}_S\mathrm{End}(M_R) = {}_S S$$

is a well-defined S-homomorphism and f is S-regular with quasi-inverse σ.
The map

$$\tau : \mathrm{Hom}_R(M, A)_T \ni h \mapsto \tau(h) := g \circ h \in \mathrm{End}(A_R)_T = T_T$$

is a well-defined T-homomorphism and f is T-regular with quasi-inverse τ.
The map

$$\sigma' : {}_{S'}\mathrm{Hom}_R(A, M) \to {}_{S'}\mathrm{End}(H_T) = {}_{S'}S' :$$

$$\forall h, x \in \mathrm{Hom}_R(M, A), \ ((h)\sigma')(x) := h \circ g \circ x$$

is a well-defined S'-homomorphism and f is S'-regular with quasi-inverse σ'.

The map

$$\tau' : \mathrm{Hom}_R(M, A)_{T'} \to \mathrm{End}(_S H)_{T'} = T'_{T'} :$$

$$\forall h, x \in \mathrm{Hom}_R(M, A), \ (x)\tau'(h) := x \circ g \circ h$$

is a well-defined T'-homomorphism and f is T'-regular with quasi-inverse τ'.
The map

$$\sigma'' : {}_{S''}\mathrm{Hom}_R(A, M) \to {}_{S''}\mathrm{End}(H_{T'}) = {}_{S''}S'' :$$

$$\forall h, x \in \mathrm{Hom}_R(M, A), \ ((h)\sigma')(x) := h \circ g \circ x$$

is a well-defined S''-homomorphism and f is S''-regular with quasi-inverse σ''.
The map

$$\tau'' : \mathrm{Hom}_R(M, A)_{T''} \to \mathrm{End}(_{S'}H)_{T''} = T''_{T''} :$$

$$\forall h, x \in \mathrm{Hom}_R(M, A), \ (x)(\tau''(h)) := x \circ g \circ h$$

is a well-defined T''-homomorphism and f is T''-regular with quasi-inverse τ''.

Proof. Clearly $\sigma, \tau, \sigma', \tau', \sigma'', \tau''$ are all additive.

For $s \in S$, we have $(sh)\sigma = shg = s(h)\sigma$. The equation $fgf = f$ now reads $(f\sigma)f = f$ showing that f is S-regular.

For $t \in T$, we have $\tau(ht) = ght = \tau(h)t$. The equation $fgf = f$ now reads $f\tau(f) = f$ showing that f is T-regular.

We first check that σ' maps into S'. In fact, let $h, x \in H$ and $t \in T$. Then $(h\sigma')(xt) = h \circ g \circ x \circ t = ((h\sigma')(x))t$. Next we check that σ' is an S'-homomorphism. Let $s' \in S'$ and $h, x \in H$. Then $g \circ x \in T$ and $((s'h)\sigma')(x) = s'(h) \circ g \circ x = s'(h \circ g \circ x) = s'((h\sigma')(x)) = (s' \circ (h\sigma'))(x)$, hence $(s'h)\sigma' = s'(h\sigma')$. Finally, $(f\sigma')f = fgf = f$.

We first check that τ' maps into T'. In fact, let $h, x \in H$ and $s \in S$. Then $(sx)(\tau')(h) = sxgh = s((x)(\tau'(h)))$. Next we check that τ' is a T'-homomorphism. Let $t' \in T'$ and $h, x \in H$. Then $x \circ g \in S$ and $(x)(\tau'(ht')) = x \circ g \circ (h)t' = ((x \circ g)h)t' = ((x)\tau'(h))t' = (x)(\tau'(h) \circ t')$, hence $\tau'(ht') = \tau'(h)t'$. Finally, $f\tau'(f) = fgf = f$.

We first check that σ'' maps into S''. In fact, let $h, x \in H$ and $t' \in T'$. Then $h \circ g \in S$ and $(h\sigma'')(xt') = h \circ g \circ (x)t' = (h \circ g \circ x)t' = (h\sigma'')(x)t'$. Next we check that σ'' is an S''-homomorphism. Let $s'' \in S''$ and $h, x \in H$. Then $g \circ x \in T'$ because $\forall s \in S, (sh)(g \circ x) = s \circ h \circ g \circ x = s \circ (h \circ g \circ x) = s((h)(g \circ x))$ and hence $((s''h)\sigma'')(x) = ((s''(h))\sigma'')(x) = s''(h) \circ g \circ x = s''(h \circ g \circ x) = s''((h\sigma'')(x)) = (s'' \circ (h\sigma''))(x)$, hence $(s''h)\sigma'' = s''(h\sigma'')$. Finally, $(f\sigma'')f \doteq fgf = f$.

We first check that τ'' maps into T''. In fact, let $h, x \in H$ and $s' \in S'$. Then $(s'x)(\tau''(h)) = (s'(x))(\tau'')(h) = s'(x) \circ g \circ h = s'(x)(g \circ h) = s'(x \circ g \circ h) = s'((x)(\tau''(h)))$. Next we check that τ'' is a T''-homomorphism. Let $t'' \in T''$ and $h, x \in H$. Then $x \circ g \in S'$ because $(s(x \circ g))(h) = s \circ x \circ g \circ h = s \circ (x \circ g \circ h) = s(x \circ g)(h)$ and $(x)(\tau''(ht'')) = (x)(\tau''((h)t'')) = x \circ g \circ (h)t'' = (x \circ g)(h)t'' = ((x \circ g) \circ h)t'' = ((x)\tau''(h))t'' = (x)(\tau''(h) \circ t'')$, hence $\tau''(ht'') = \tau''(h)t''$. Finally, $f\tau''(f) = fgf = f$. $\qquad\square$

It is not clear how the other types of regularity are related.

The following example shows that regular and S-regular need not be equivalent.

Example 2.2. There exist $\mathbb{Z}$-modules A and M such that $\operatorname{Hom}_R(A, M)$ contains no nonzero regular elements while $\operatorname{Hom}_R(A, M)$ contains nonzero $\operatorname{End}(A_R)$-regular elements.

Proof. Let p be a prime number, $A = a\mathbb{Z}$, $\operatorname{ord}(a) = p$, and $B = b\mathbb{Z}$, $\operatorname{ord}(b) = p^2$. Suppose that $f \in \operatorname{Hom}(A, M)$ is regular. Then $M = \operatorname{Im}(f) \oplus U$ for some subgroup U. Since M is indecomposable and $|\operatorname{Im}(f)| \leq |a\mathbb{Z}| = p$, it is not possible that $\operatorname{Im}(f) = M$, so $\operatorname{Im}(f) = 0$, and $f = 0$.

Note that $T := \operatorname{End}(a\mathbb{Z}) \cong \mathbb{Z}/p\mathbb{Z}$ and $S := \operatorname{End}(b\mathbb{Z}) \cong \mathbb{Z}/p^2\mathbb{Z}$. Let $f \in \operatorname{Hom}(a\mathbb{Z}, b\mathbb{Z})$ be the map with $f(a) = pb$ and define for $h \in \operatorname{Hom}(a\mathbb{Z}, b\mathbb{Z})$ with $h(a) = pbk$ the functional $\varphi : \operatorname{Hom}(a\mathbb{Z}, b\mathbb{Z}) \ni h \mapsto k + p\mathbb{Z} \in \mathbb{Z}/p\mathbb{Z} = \operatorname{End}(a\mathbb{Z})$. Then $\varphi(f) = 1$ and $f\varphi(f) = f$ so that f is a nonzero T-regular element. $\square$

Question 2.3. When does f regular follow from the fact that f is S, T, S', T'-regular?

We will make use of a lemma of Ware ([35, Lemma 2.2]) that is a generalization of a result of Chase ([5, Proposition 2.1], and used by Zelmanowitz ([37, Lemma 2.4]). We include it with proof.

Lemma 2.4. *Let $_S P$ be a projective S-module and K a submodule of P. Then P/K is flat if and only if for every $x \in K$ there is a homomorphism $\alpha_x : P \to K$ with $x\alpha_x = x$.*

Proof. (a) We first assume that P is free on a basis $\{b_i \mid i \in I\}$.

Assume that P/K is flat, let $x \in K$, and write $x = \sum_{i \in I} s_i b_i$, where $s_i \in S$ and all but finitely many of the s_i are equal to 0. Let $I_x := \sum_{i \in I} s_i S$. Then I_x is a finitely generated right ideal of S. By Proposition I.2.4 we have $K \cap I_x P = I_x K$ which contains x and hence $x = \sum_{i \in I} t_i k_i$ where $k_i \in K$, $t_i \in I_x$, and all but finitely many of the t_i are equal to 0. Define $\alpha_x : P \to K$ by stipulating that $b_i \alpha_x = k_i$ if $s_i \neq 0$ and $b_i \alpha_x = 0$ otherwise. Then $x\alpha_x = \left(\sum_{i \in I} s_i b_i \right) \alpha_x = \sum_{i \in I} t_i k_i = x$.

Conversely, assume that for every $x \in K$ there is $\alpha_x : P \to K$ such that $x\alpha_x = x$. Let U be a right ideal of S and let $x \in K \cap UP$ be given. Then $x = \sum_{i \in I} u_i b_i$, where $u_i \in U$. Applying α_x we get $x = x\alpha_x = \left(\sum_{i \in I} u_i b_i \right) \alpha_x = \sum_{i \in I} u_i(b_i \alpha_x) \in UK$, showing that $K \cap UP = UK$. By Proposition I.2.4 we see that P/K is flat.

(b) Now we consider the general case where P is projective but not necessarily free. Then there exists a free module F such that $F = P \oplus L$ for some module L.

Suppose that P/K is flat and let $x \in K$. Then $P/K \oplus Q$ is flat hence, by (a), there exists $g : F \to K$ such that $xg = x$. Let $\alpha_x := g \upharpoonright_P$. Then $\alpha_x : P \to K$ with $x\alpha_x = x$.

Assume that for every $x \in K$ there is a homomorphism $\alpha_x : P \to K$ with $x\alpha_x = x$. Let $x \in K$. Then there exists $\alpha_x : P \to K$ such that $x\alpha_x = x$. Extend α_x to $g : F \to K$ by defining $Lg = 0$. Then $xg = x$ and by (a) $(P \oplus L)/K$ is flat. Therefore P/K is flat. $\square$

Proposition 2.5. *Every epimorphic image of* $_S \text{Reg}(A, M)$ *is* S-*flat, and every epimorphic image of* $\text{Reg}(A, M)_T$ *is* T-*flat.*

Proof. Suppose that K is an S-submodule of $\text{Reg}(A, M)$. We will show that $\text{Reg}(A, M)/K$ is flat. Since direct limits of flat modules are flat, it suffices to show that every finitely generated submodule of $\text{Reg}(A, M)/K$ is flat. A finitely generated submodule of $\text{Reg}(A, M)/K$ is of the form $(N + K)/K$ where N is a finitely generated submodule of $\text{Reg}(A, M)$. The module N is regular and projective by Theorem II.4.6. We wish to apply Lemma 2.4 to conclude that $(N + K)/K \cong N/(N \cap K)$ is flat. We have that N is a projective module and we need to show that for every $x \in N \cap K$ there is an S-homomorphism $\alpha_x : N \to N \cap K$ such that $x\alpha_x = x$. Let $x \in N \cap K$. As N is regular as a subset of $\text{Hom}_R(A, M)$, it is also S-regular (see Lemma 2.1), and hence there exists $\varphi_x : N \to S$ such that $(x\varphi_x)x = x$. Define $\alpha_x : N \ni n \mapsto (n\varphi_x)x$ which is an S-homomorphism. Then $x\alpha_x = (x\varphi_x)x = x$. By Lemma 2.4 we obtain that $(N + K)/K \cong N/(N \cap K)$ is flat. $\square$

We now have numerous immediate corollaries to our results on regularity for modules.

Corollary 2.6 (Characterization of Regularity). *Let* P *be any one of the rings* S, S', *or* S''. *Let* $f \in {}_P \text{Hom}_R(A, M)$. *Then the following statements are equivalent.*

1) f *is* P-*regular.*

2) $\text{Ann}_P(f) \subseteq^\oplus P$ *and* $Pf \subseteq^\oplus \text{Hom}_R(A, M)$.

3) Pf *is a* P-*projective direct summand of* $\text{Hom}_R(A, M)$.

Let P *be any one of the rings* T, T', *or* T''. *Let* $f \in \text{Hom}_R(A, M)_P$. *Then the following statements are equivalent.*

1) f *is* P-*regular.*

2) $\text{Ann}_P(f) \subseteq^\oplus P$ *and* $fP \subseteq^\oplus \text{Hom}_R(A, M)$.

3) fP *is a* P-*projective direct summand of* $\text{Hom}_R(A, M)$.

Proof. Theorem IV.1.4. $\square$

This also says that all elements of $\text{Reg}(A, M)$ are S-, T-, S'-, T'-, S''- and T''-regular. Later we will come back to this fact.

The next and some later results we only formulate for $_S \text{Hom}_R(A, M)$ but there are five more results corresponding to the other five module structures on H.

Proposition 2.7. *The following statements hold.*

1) *Let $f \in H = \mathrm{Hom}_R(A, M)$ be S-regular. Then there exists a quasi-inverse $\varphi \in \mathrm{Hom}_S(H, S)$ of f that is also regular and has quasi-inverse f, i.e., in addition to $f \circ (f)\varphi = f$ we have that $(f)\varphi \circ \varphi = \varphi$.*

2) *Suppose that $f \in H$ and $\varphi \in H^* = \mathrm{Hom}_S(H, S)$ are such that $f \circ (f)\varphi = f$ and $\varphi \circ (f)\varphi = \varphi$. Then*

$$S = \mathrm{Ann}_S(f) \oplus \mathrm{Im}(\varphi), \quad and \quad \mathrm{Hom}_R(A, M) = Sf \oplus \mathrm{Ker}(\varphi).$$

Proof. Lemma 2.1. □

Proposition 2.8. *Suppose that $f \in \mathrm{Hom}_R(A, M)$ is S-regular and $T' = \mathrm{End}(_S H)$ is commutative. Then f has a unique quasi-inverse.*

Proof. Proposition IV.2.5. □

Corollary 2.9. *Let $f \in \mathrm{Hom}_R(A, M)$ be S-regular with quasi-inverse $\varphi \in \mathrm{Hom}_S(H, S)$ and suppose that S is commutative. Then f is the unique quasi-inverse of $(f)\varphi \circ \varphi$.*

Proof. Proposition IV.2.6. □

3 Largest Regular Submodules

The existence of various largest regular submodules is an immediate consequence of Theorem IV.1.8.

Definition 3.1. Let $H = \mathrm{Hom}_R(A, M)$, S, T, etc. be as before. We define

1) $\mathrm{Reg}(_S H) := \{f \in H \mid SfT' \text{ is } S\text{-regular}\}$.

2) $\mathrm{Reg}(H_T) := \{f \in H \mid S'fT \text{ is } T\text{-regular}\}$.

3) $\mathrm{Reg}(_{S'} H) := \{f \in H \mid S'fT'' \text{ is } S'\text{-regular}\}$.

4) $\mathrm{Reg}(H_{T'}) := \{f \in H \mid S''fT' \text{ is } T'\text{-regular}\}$.

5) $\mathrm{Reg}(_{S''} H) := \{f \in H \mid S''fT' \text{ is } S''\text{-regular}\}$.

6) $\mathrm{Reg}(H_{T''}) := \{f \in H \mid S'fT'' \text{ is } T''\text{-regular}\}$.

Here it is used that $S''' = S'$ and $T''' = T'$.

Theorem 3.2.

1) $\mathrm{Reg}(_S H)$ *is the largest S-regular S-T'-submodule of $_S H_{T'}$.*

2) $\mathrm{Reg}(H_T)$ *is the largest T-regular S'-T-submodule of $_{S'} H_T$.*

3) $\mathrm{Reg}(_{S'} H)$ *is the largest S'-regular S'-T''-submodule of $_{S'} H_{T''}$.*

4) $\mathrm{Reg}(H_{T'})$ is the largest T'-regular S''-T'-submodule of $_{S''}H_{T'}$.

5) $\mathrm{Reg}(_{S''}H)$ is the largest S''-regular S''-T'-submodule of $_{S''}H_{T'}$.

6) $\mathrm{Reg}(H_{T''})$ is the largest T''-regular S'-T''-submodule of $_{S'}H_{T''}$.

Proof. Theorem IV.1.8. $\qquad\qquad\qquad\qquad\qquad\qquad\qquad\qquad\qquad$ □

We have already mentioned that f regular implies that f is S-, T-, S'-, T'-, S''-, and T''-regular. Hence $\mathrm{Reg}(A, M)$ is S-, T-, etc. regular. But we do not know whether $\mathrm{Reg}(A, M)$ is contained in $\mathrm{Reg}(_S H)$, $\mathrm{Reg}(H_T)$, $\mathrm{Reg}(_{S'}H)$, $\mathrm{Reg}(H_{T'})$, $\mathrm{Reg}(_{S''}H)$, or $\mathrm{Reg}(H_{T''})$. We have that $f \in \mathrm{Reg}(A, M)$ if and only if SfT is regular; however this does not necessarily mean that SfT' is S-regular. Similarly for the other cases.

Question 3.3. For which pairs A, M does SfT regular imply SfT' is S-regular? Same question for other combinations of rings.

The modules $\mathrm{Reg}(_S H)$, $\mathrm{Reg}(H_T)$, etc. have the properties that $\mathrm{Reg}(M_R)$ has for a general module M_R and which were derived from the properties of $\mathrm{Reg}(A, M)$ with $A = R$. We list these properties for $\mathrm{Reg}(_S H)$ only; it is easy to formulate them for $\mathrm{Reg}(_S H)$, $\mathrm{Reg}(H_T)$ and the other module structures on H.

Theorem 3.4. *Let N be a finitely generated S'-submodule of $\mathrm{Reg}(H_T)$. Then N is an S'-projective direct summand of H. Furthermore, N is the direct sum of finitely many cyclic projective submodules, each of which is isomorphic to a left ideal of S'. The same is true for submodules of $\mathrm{Reg}(H_T)$ considered as a right T-module.*

Proof. Theorem IV.3.3. $\qquad\qquad\qquad\qquad\qquad\qquad\qquad\qquad\qquad\quad$ □

Theorem 3.5. *Suppose that $\mathrm{Reg}(H_T)$ contains no infinite direct sums of S'-submodules. Then $\mathrm{Reg}(H_T)$ is the direct sum of finitely many simple projective S'-modules, $\mathrm{Hom}_R(A, M) = \mathrm{Reg}(H_T) \oplus U$ for some S'-submodule U and U contains no nonzero regular S'-T-submodule. Every cyclic S'-submodule of $\mathrm{Reg}(H_T)$ is isomorphic to a left ideal of S' that is a direct summand of S'.*

Analogous results hold for the other module structures on $\mathrm{Hom}_R(A, M)$.

Proof. Theorem IV.3.4. $\qquad\qquad\qquad\qquad\qquad\qquad\qquad\qquad\qquad\quad$ □

Theorem 3.6. *Suppose that S' is left perfect. Then $\mathrm{Reg}(H_T)$ is S'-projective. If T is right perfect, then $\mathrm{Reg}(H_T)$ is T-projective.*

Proof. Theorem IV.3.7. $\qquad\qquad\qquad\qquad\qquad\qquad\qquad\qquad\qquad\quad$ □

4   The Transfer Rule for S-Regularity

Here also we repeat the transfer rule of the plain regular case but we enter situations where regularity was not applied before.

Let f be S-regular and $f \sigma f = f$ with $\sigma \in \mathrm{Hom}_S(H, S)$. Then we get the idempotent

$$e := f\sigma = e^2 \in S, \text{ and } ef = f.$$

Further $\sigma(f\sigma) \in \mathrm{Hom}_S(H, S)$ where

$$\sigma(f\sigma) : H \ni h \mapsto (h\sigma)(f\sigma) \in S.$$

As before, we call (f, σ) a regular pair and define

$$\mathrm{trf}(f, \sigma) = (\sigma f \sigma, f)$$

where $\sigma f \sigma = \sigma e \in \mathrm{Hom}_S(H, S)$. Further, $(\sigma f \sigma, f)$ is again a regular pair because

$$\sigma f \sigma f \sigma f \sigma = \sigma e^3 = \sigma e = \sigma f \sigma$$

with $\sigma f \sigma = \sigma e \in \mathrm{Hom}_S(H, S)$. It follows further that

$$\mathrm{trf}(\sigma f \sigma, f) = (f \sigma f \sigma f, \sigma f \sigma) = (f, \sigma e).$$

Corollary 4.1. *If* trf *is applied to all S-regular pairs (f, σ) with $f \in \mathrm{Hom}_R(A, M)$, then the set of all entries $\sigma f \sigma$ in* $\mathrm{trf}(f, \sigma) = (\sigma f \sigma, f)$ *is the set of all S-regular elements in $\mathrm{Hom}_S(H, S)$.*

Proof. Same as that of Theorem II.5.3. $\qquad\qquad\qquad\qquad\square$.

Chapter VI

Relative Regularity: U-Regularity and Semiregularity

1 U-Regularity; Definition and Existence of U-$\mathrm{Reg}(A, M)$

The paper [19] on U-regularity was the first step in the study of regular substructures of Hom. We give here the main properties of U-regularity.

Definition 1.1.

1) Let U be an S-T-submodule of $_SH_T = {}_S\mathrm{Hom}(A, M)_T$. Then $f \in H$ is called U-**regular** if and only if there exists $g \in \mathrm{Hom}_R(M, A)$ such that

$$fgf - f \in U. \tag{1}$$

2) U-$\mathrm{Reg}(A, M) := \{f \in H \mid SfT \text{ is } U\text{-regular}\}$ where SfT is the S-T-submodule of H generated by f and "SfT is U-regular" means that every element of SfT is U-regular.

We add some comments that will serve to familiarize the reader with the new notions.

1) For $U = 0$, U-regular = regular.

2) $U \subseteq$ U-$\mathrm{Reg}(A, M)$ because U is an S-T-submodule, and for $u \in U$ we have $u \cdot 0 \cdot u - u = -u \in U$, hence u is in U-$\mathrm{Reg}(A, M)$.

3) If U_1 and U_2 are S-T-submodules of H, and $U_1 \subseteq U_2$, then U_1-$\mathrm{Reg}(A, M) \subseteq U_2$-$\mathrm{Reg}(A, M)$.

4) If $f \in H$ is U-regular with (1) and $v \in U$, then also $f + v$ is U-regular, since, by (1), setting $u := fgf - f \in U$, we get $(f + v)g(f + v) - (f + v) = u_1 \in U$ where $u_1 = u + v + fgv + vgf$. Here we used the fact that $fg \in S$, $gf \in T$ and that U is an S-T-module. This implies also that $S(f + v)T$ is U-regular if and only if SfT is U-regular.

For U-regularity we have the same important theorem that we proved earlier for other regularities.

Theorem 1.2. *The set* U-$\mathrm{Reg}(A, M)$ *is the unique largest* U*-regular* S-T*-submodule of* $\mathrm{Hom}_R(A, M)$.

The proof is more or less the same as in the previous case (namely Theorem II.4.3), yet we will give it in detail to reinforce the arguments. For the proof we need the following lemma.

Lemma 1.3. *Let* U *be an* S-T*-submodule of* $\mathrm{Hom}_R(A, M)$, *let* $f \in \mathrm{Hom}_R(A, M)$ *and let* $g \in \mathrm{Hom}_R(M, A)$. *If* $f - fgf$ *is* U*-regular, then* f *is* U*-regular.*

Proof. Since $f - fgf$ is U-regular, there exists $h \in \mathrm{Hom}_R(M, A)$ such that

$$(f - fgf)h(f - fgf) - (f - fgf) \in U. \tag{2}$$

Let

$$k := h - gfh - hfg + gfhfg + g \in \mathrm{Hom}_R(M, A)$$

with h from (2). An easy computation shows that $fkf - f \in U$. In fact,

$$\begin{aligned} fkf &= fhf - fgfhf - fhfgf + fgfhfgf + fgf \\ &= (f - fgf)h(f - fgf) + fgf \end{aligned}$$

and by (2)

$$f - fkf = (f - fgf) - (f - fgf)h(f - fgf) \in U. \qquad \square$$

Proof of Theorem 1.2. This proof is achieved in four steps.

(a) Let $f \in H$ and assume that SfT is U-regular. If $\varphi \in SfT$, then $S\varphi T \subseteq SfT$, hence also $S\varphi T$ is U-regular and $\varphi \in U$-$\mathrm{Reg}(A, M)$. Since φ was an arbitrary element of SfT, it follows that $SfT \subseteq U$-$\mathrm{Reg}(A, M)$.

(b) Since U-$\mathrm{Reg}(A, M)$ is the sum of modules of the form SfT, it is closed under left multiplication by S and right multiplication by T.

(c) We show now that U-$\mathrm{Reg}(A, M)$ is closed under addition. Let $f_1, f_2 \in U$-$\mathrm{Reg}(A, M)$. We have to show that $S(f_1 + f_2)T$ is U-regular. An element of $S(f_1 + f_2)T$ is of the form

$$\sum_{i=1}^n s_i(f_1 + f_2)t_i = \sum_{i=1}^n s_i f_1 t_i + \sum_{i=1}^n s_i f_2 t_i.$$

Set $\varphi_1 = \sum_{i=1}^n s_i f_1 t_i$ and $\varphi_2 = \sum_{i=1}^n s_i f_2 t_i$. Then $\varphi_1 \in Sf_1T$ is U-regular and there exists $g \in \mathrm{Hom}_R(M, A)$ such that $\varphi_1 g \varphi_1 - \varphi_1 \in U$. Now consider

$$(\varphi_1 + \varphi_2) - (\varphi_1 + \varphi_2)g(\varphi_1 + \varphi_2) = (\varphi_1 - \varphi_1 g \varphi_1) + (\varphi_2 - \varphi_1 g \varphi_2 - \varphi_2 g \varphi_1 - \varphi_2 g \varphi_2). \tag{3}$$

Here $\varphi_1 g \varphi_1 - \varphi_1 \in U$ and since $\varphi_1 g \in S$, $g\varphi_1, \varphi_2 g \in T$, the term $\varphi_2 - \varphi_1 g \varphi_2 - \varphi_2 g \varphi_1 - \varphi_2 g \varphi_2$ in (3) is in $S\varphi_2 T$. We now use that $S\varphi_2 T$ is U-regular which is true by (a). Therefore the element in (3) is U-regular and we can apply Lemma 1.3 with $\varphi_1 + \varphi_2$ in place of f, and obtain that $\varphi_1 + \varphi_2$ is U-regular. Since $\varphi_1 + \varphi_2$ was an arbitrary element of $S(f_1 + f_2)T$, this submodule is U-regular. So $S(f_1 + f_2)T \subseteq$ U-Reg(A, M) and in particular, $f_1 + f_2 \in$ U-Reg(A, M).

(d) Finally we show that U-Reg(A, M) is the largest U-regular S-T-submodule of H. Assume that Λ is a U-regular S-T-submodule of H and $f \in \Lambda$. Then SfT is a U-regular S-T-submodule of H and by definition of U-Reg(A, M) we have $SfT \subseteq$ U-Reg(A, M), in particular $f \in$ U-Reg(A, M). This means that $\Lambda \subseteq$ U-Reg(A, M) and U-Reg(A, M) is the largest regular S-T-submodule of H. $\quad\square$

In the following A and M will be fixed and we will therefore write U-Reg instead of U-Reg(A, M). As U-Reg is again an S-T submodule of H, U-Reg can be used in place of U in the definition above, i.e., we may consider (U-Reg)- Reg.

Theorem 1.4. *Let U be an S-T-submodule of $H = {}_S\mathrm{Hom}(A, M)_T$. Then*

$$(\text{U-Reg}(A, M))\text{-}\mathrm{Reg}(A, M) = \text{U-Reg}(A, M).$$

Proof. Since $U \subseteq$ U-Reg, it follows that

$$\text{U-Reg} \subseteq (\text{U-Reg})\text{-}\mathrm{Reg}.$$

We have to prove the converse inclusion. Let $f \in (\text{U-Reg})\text{-}\mathrm{Reg}$. Then there exist $g \in \mathrm{Hom}_R(M, A)$ and $w \in$ U-Reg such that

$$fgf - f = w. \tag{4}$$

Since $w \in$ U-Reg, there exists $h \in \mathrm{Hom}_R(M, A)$ with

$$whw - w = u \in U. \tag{5}$$

It follows from (4) that

$$w = f(gf - 1_T) = (fg - 1_S)f \tag{6}$$

and using (4), (5) and (6) together we obtain

$$
\begin{aligned}
f &= fgf - w = fgf - whw + u \\
&= fgf - f(gf - 1_T)h(fg - 1_S)f + u \\
&= f[g - (gf - 1_T)h(fg - 1_S)]f + u.
\end{aligned}
$$

The element in the bracket is in $\mathrm{Hom}_R(M, A)$. Hence the equation fits the definition for f to be U-regular. We have now established that every (U-Reg)-regular element is U-regular. $\quad\square$

2 U-Regularity in Modules

As before we can use the isomorphism $\rho : \mathrm{Hom}_R(R, M) \ni f \mapsto f(1) \in M$ to specialize U-regularity. As always $S := \mathrm{End}(M_R)$ and $R = \mathrm{End}(R_R)$. Then $\mathrm{Hom}_R(R, M)$ and M are naturally S-R-modules and ρ becomes a bimodule isomorphism. Let U be an S-R-submodule of M and $U' = \rho^{-1}(U)$. Then $f \in \mathrm{Hom}_R(R, M)$ is U'-regular if and only if there exists $\varphi \in M^* := \mathrm{Hom}_R(M, R)$ such that $f\varphi f - f \in U'$. Setting $m = f(1)$ and applying ρ we obtain that $m\varphi(m) - m \in U$. We define U-regularity in M in such a way that m is U-regular in M if and only if $\rho^{-1}(m)$ is $\rho^{-1}(U)$-regular in $\mathrm{Hom}_R(R, M)$.

Definition 2.1.

1) Let U be an S-R-submodule of M. Then $m \in M$ is called U-**regular** if and only if there exists $\varphi \in \mathrm{Hom}_R(M, R)$ such that

$$m\varphi(m) - m \in U.$$

2) U-Reg$(M_R) := \{m \in M \mid SmR \text{ is } U\text{-regular}\}$ where SmR is the S-R-submodule of M generated by m and "SmR is U-regular" means that every element of SmR is U-regular.

As in the general case we have the following facts.

1) For $U = 0$, U-regular = regular.

2) $U \subseteq$ U-Reg(M_R).

3) If U_1 and U_2 are S-R-submodules of M, and $U_1 \subseteq U_2$, then U_1-Reg$(M_R) \subseteq U_2$-Reg(M_R).

4) If $m \in M$ is U-regular and $v \in U$, then also $m + v$ is U-regular and $S(m+v)R$ is U-regular if and only if SmR is U-regular.

Theorem 2.2. *The set* U-Reg(M_R) *is the unique largest U-regular S-R-submodule of M.*

Theorem 2.3. *Let U be an S-R-submodule of $_SM_R$. Then*

$$(\text{U-Reg}(M_R))\text{-}\mathrm{Reg}(M_R) = \text{U-Reg}(M_R).$$

We will now compare U-regularity with regularity in M/U.

Proposition 2.4. *Let U be an S-R-submodule of M_R, and let*

$$- : M \ni m \mapsto \overline{m} := m + U \in M/U = \overline{M}$$

be the natural R-epimorphism. Let $m \in M$.

1) *If $\overline{m}$ is regular in $\overline{M}$, then m is U-regular in M.*

2) *If* $\operatorname{Hom}_R(U, R) = 0$ *and* m *is regular in* M, *then* $\overline{m}$ *is* U-*regular in* $\overline{M}$.

Proof. 1) Note that any element of $\overline{M}$ is of the form $\overline{m}$ for some $m \in M$. Let $\overline{m}$ be regular in $\overline{M}$. Then there is $\psi \in \operatorname{Hom}_R(\overline{M}, R)$ such that $\overline{m}(\psi(\overline{m})) = \overline{m}$ which says that $m((\psi \circ {}^{-})(m)) - m \in U$ where $\psi \circ {}^{-} \in \operatorname{Hom}_R(M, R)$.

2) Assume that $\operatorname{Hom}_R(U, R) = 0$, let m be U-regular in M, and let $\varphi \in \operatorname{Hom}(M, R)$ be such that $m\varphi(m) - m \in U$. The hypothesis $\operatorname{Hom}_R(U, R) = 0$ implies that φ factors through ${}^{-} \in \operatorname{Hom}_R(M/U, R)$, i.e., there is $\psi \in \operatorname{Hom}_R(M/U, R)$ such that $\varphi = \psi \circ {}^{-}$. We conclude that $\overline{m}(\psi(\overline{m})) - \overline{m} = \overline{m\phi(m) - m} = 0$. $\qquad\square$

We now look at some examples.

Example 2.5. Let $M_{\mathbb{Z}}$ be an abelian group. Then:

1) $0 \neq m \in M$ is regular if and only if $M = m\mathbb{Z} \oplus N$ for some subgroup N and $m\mathbb{Z} \cong \mathbb{Z}$. However $\operatorname{Reg}(M_{\mathbb{Z}}) = 0$.

2) Let p a prime, and suppose that $\operatorname{Hom}(pM, \mathbb{Z}) = 0$. Then pM is a fully invariant subgroup of M, $(M/pM)_{\mathbb{Z}}$ contains no nonzero regular elements although M/pM is regular as a module over the field $\mathbb{Z}/p\mathbb{Z}$. We have $pM - \operatorname{Reg}(M) = pM$.

3) $\mathbf{t}(M) - \operatorname{Reg}(M_R) = \mathbf{t}(M)$

Proof. 1) Suppose that $0 \neq m \in M$ is regular. Then there is $\varphi \in \operatorname{Hom}(M, \mathbb{Z})$ such that $m\varphi(m) = m \neq 0$. Hence $\varphi \neq 0$. By Theorem IV.1.3 we get $\mathbb{Z} = \operatorname{Ann}_{\mathbb{Z}}(m) \oplus \varphi(m)\mathbb{Z}$ and hence $\mathbb{Z} = \varphi(m)\mathbb{Z}$, $M = m\mathbb{Z} \oplus \operatorname{Ker}(\varphi)$, and $m\mathbb{Z} \cong \varphi(m)\mathbb{Z} = \mathbb{Z}$. Conversely, if $M = m\mathbb{Z} \oplus N$ and $m\mathbb{Z} \cong \mathbb{Z}$, then the map $\varphi \in \operatorname{Hom}(M, \mathbb{Z})$ with $\varphi(mk) = k$ and $\varphi(N) = 0$ clearly is a quasi-inverse for m. By Theorem 1.2 we have $\operatorname{Reg}(M_{\mathbb{Z}}) = 0$.

2) We observe first that pM is a fully invariant submodule of M and hence is an S-$\mathbb{Z}$-submodule of M where $S := \operatorname{End}(M)$. Observe that M/pM is a semisimple $\mathbb{Z}/p\mathbb{Z}$-module, and hence is regular. However as an abelian group it contains no nonzero regular elements whatsoever. Thus if $m \in M$ is pM-regular, then $\overline{m} = 0$, i.e., $m \in pM$. This shows that $pM - \operatorname{Reg}(M_{\mathbb{Z}}) = pM$.

3) Again $\mathbf{t}(M)$ is fully invariant in M, and $\operatorname{Hom}(\mathbf{t}(M), \mathbb{Z}) = 0$. In this case $(M/\mathbf{t}(M))_{\mathbb{Z}}$ may well contain nonzero regular elements, but if $\overline{m} \neq 0$ is such an element, then $\overline{mk}$ where $0 \neq k \in \mathbb{Z}$ cannot be regular. So km is not $\mathbf{t}(M)$-regular in M while $k \in \operatorname{End}(M)$. The claim follows. $\qquad\square$

3   Semiregularity for Modules

Another form of regularity is semiregularity. This was introduced and studied for modules by W.K. Nicholson ([26]).

Definition 3.1. (Nicholson) Let M_R be an R-module. An element $m \in M$ is **semiregular** if there exists a regular element $n \in M$ such that $n - m \in \operatorname{Rad}(M_R)$.

Nicholson supplies the following equivalent characterizations of semiregularity. To state them, we need the notion of "lying over".

Definition 3.2. A submodule $C \subseteq M$ **lies over a summand of** M if and only if $M = A \oplus B$ with $A \subseteq C$ and $C \cap B$ is small.

Proposition 3.3 ([26, Proposition 1.3]). *Let M be a right R-module and $m \in M$. The following conditions on m are equivalent.*

1) *m is semiregular.*

2) *There is $n \in M$ such that n is regular, and $n - m \in \mathrm{Rad}(M)$.*

3) *There is $e^2 = e \in \mathrm{End}(M_R)$ such that $e(M) \subseteq mR$, $e(M)$ is projective and $m - e(m) \in \mathrm{Rad}(M)$.*

4) *mR lies over a projective direct summand of M.*

5) *There exists $\varphi \in M^* := \mathrm{Hom}_R(M, R)$ such that $\varphi(m)^2 = \varphi(m)$ and $m - m\varphi(m) \in \mathrm{Rad}(M)$.*

6) *There is $n \in mR$ such that n is regular, $n - m \in \mathrm{Rad}(M)$, and $mR = nR \oplus (m - n)R$.*

Proof. 1) $\Leftrightarrow$ 2) is the definition.

2) $\Rightarrow$ 3): By hypothesis we have a regular element $n \in M$ such that $n - m \in \mathrm{Rad}(M)$. As n is regular we further have $\varphi \in M^*$ such that

$$n\varphi(n) = n, \text{ so also } \varphi(n)^2 = \varphi(n). \tag{7}$$

Assume for the moment that $n \in mR$. Let $e : M \ni x \mapsto e(x) = n\varphi(x) \in M$. Then $e^2 = (n\varphi)(n\varphi) = n\varphi(n)\varphi(n) = n\varphi(n) = e$, with $e(M) \subseteq nR \subseteq mR$. As n is regular, by Theorem IV.1.3, the submodule nR is a projective direct summand of M and hence $e(M)$, being a direct summand of nR, is also projective. Finally, $m - e(m) = m - n\varphi(m) = m - n + n - n\varphi(m) = (m - n) + n\varphi(n) - n\varphi(m) = (m - n) + e(n - m) \in \mathrm{Rad}(M)$.

We now drop the assumption that $n \in mR$ and show that $m\varphi(n)$ is regular with $m - m\varphi(n) \in \mathrm{Rad}(M)$. Therefore $m\varphi(n) \in mR$ can be used in place of n showing that there was no loss of generality in assuming $n \in mR$.

Note that $\varphi(n) - \varphi(m) = \varphi(n - m) \in \mathrm{Rad}(R)$, and therefore there is $s \in R$ such that $s(1 - \varphi(n) + \varphi(m)) = 1$. Then, using (7),

$$\varphi(n) = s(1 - \varphi(n) + \varphi(m))\varphi(n) = s\varphi(m)\varphi(n). \tag{8}$$

We have $s\varphi \in M^*$ and find that

$$(m\varphi(n))(s\varphi)(m\varphi(n)) = (m\varphi(n))(s\varphi(m)\varphi(n))$$
$$\overset{(8)}{=} (m\varphi(n))\varphi(n) \overset{(7)}{=} m\varphi(n),$$

showing that $m\varphi(n)$ is regular with quasi-inverse $s\varphi$. Furthermore, $m - m\varphi(n) = m - n + n - m\varphi(n) = (m-n) + n\varphi(n) - m\varphi(n) = (m-n) + (n-m)\varphi(n) \in \operatorname{Rad}(M)$.

3) $\Rightarrow$ 4): Obvious.

4) $\Rightarrow$ 5): By hypothesis there is a decomposition $M = K \oplus L$ such that $K \subseteq mR$, K is projective, and $L \cap mR$ is small. Hence $mR = K \oplus (L \cap mR)$ and it follows that $K = kR$ where $m = k + \ell$ for some $\ell \in L \cap mR$. As K is projective, by the Basis Lemma applied to the generating set $\{k\}$, we have $\varphi \in K^*$ such that for every $x \in K$, $x = k\varphi(x)$. Extend φ to all of M by stipulating that $\varphi(L) = 0$. Then $\varphi \in M^*$ and $\varphi(m)^2 = \varphi(k)^2 = \varphi(k\varphi(k)) = \varphi(k) = \varphi(m)$. Finally, $m - m\varphi(m) = k + \ell - m\varphi(k) = k + \ell - k\varphi(k) - \ell\varphi(k) = \ell - \ell\varphi(k) \in L \cap mR \subseteq \operatorname{Rad}(M)$.

5) $\Rightarrow$ 6) By hypothesis we have $\varphi \in M^*$, $\varphi(m)^2 = \varphi(m)$, and $m - m\varphi(m) \in \operatorname{Rad}(M)$. Set $n := m\varphi(m)$. Then $m - n \in \operatorname{Rad}(M)$, and $mR = m\varphi(m)R \oplus m(1 - \varphi(m))R = nR \oplus (m-n)R$. Hence $(m-n)R$ is cyclic and contained in the radical, so small. It remains to show that n is regular. As $n\varphi(n) = m\varphi(m)\varphi(m\varphi(m)) = m\varphi(m)\varphi(m)\varphi(m) = m\varphi(m) = n$ this is indeed the case.

6) $\Rightarrow$ 2) is again obvious. $\square$

An important characterization of semiregular modules, also due to Nicholson, follows.

Theorem 3.4 ([26, Theorem 1.6]). *For a module M_R the following statements are equivalent.*

1) *M is semiregular.*

2) *For every finitely generated submodule N of M, there is $e^2 = e \in \operatorname{End}(M_R)$ such that $e(M) \subseteq N$, $e(M)$ is projective and $N(1 - e) \subseteq \operatorname{Rad}(M)$.*

3) *Every finitely generated submodule N of M lies over a projective direct summand of M.*

Proof. 1) $\Rightarrow$ 2): We induct on the number n of generators of N. For $n = 1$ the claim follows from Proposition 3.3.3). Suppose that $n > 1$ and $N = a_1 R \oplus \cdots \oplus a_n R$. Choose $e^2 = e \in \operatorname{End}(M_R)$ such that $e(M) \subseteq a_n R$, $e(M)$ is projective, and $(1 - e)(a_n R) \subseteq \operatorname{Rad}(M)$. Set $K := (1 - e)(a_1)R \oplus \cdots \oplus (1 - e)(a_{n-1})R$, and, by induction hypothesis, choose $d^2 = d \in \operatorname{End}(M_R)$ such that $d(M) \subseteq K$, $d(M)$ is projective and $(1 - d)(K) \subseteq \operatorname{Rad}(M)$. Then $e(K) = 0$, hence $(ed)(M) = 0$ which means that $ed = 0$. We have $M = e(M) \oplus (1 - e)(M) = d(M) \oplus (1 - d)(M)$ and $ed = 0$. Now set $k := e + d - de$. Then

$$
\begin{aligned}
k^2 &= (e + d - de)(e + d - de) = e^2 + de + d^2 - d^2 e - de^2 \\
&= e + de + d - de - de = e + d - de = k
\end{aligned}
$$

and we easily find that

$$
ke = e^2 + de - de^2 = e, \quad \text{and} \quad kd = d^2 = d.
$$

Further, $N = K + a_n R \supseteq d(M) + e(M)$ and $d(M) + e(M) = d(M) \oplus e(M)$ because $ed = 0$. Also $k(M) = (e + d - de)(M) \subseteq e(M) + d(M) = ke(M) + kd(M) \subseteq k(M)$, so $k(M) = d(M) \oplus e(M)$ is projective. Finally, $(1 - k)(N) = (1 - d - e + de)(N) = (1 - d)(1 - e)(N) \subseteq \mathrm{Rad}(M)$.

2) $\Rightarrow$ 3): We have $M = e(M) \oplus (1 - e)(M)$ and $(1 - e)(N) = N \cap (1 - e)(M) \subseteq \mathrm{Rad}(M)$. Since $(1 - e)(N)$ is finitely generated, it is in fact small. Hence N lies over a projective summand of M.

3) $\Rightarrow$ 1): By hypothesis and Proposition 3.3 every element of M is semiregular. $\qquad\square$

Zelmanowitz studied the class of regular modules, and Nicholson was interested in the wider class of semiregular modules. We list some of Nicholson's more striking results without proof.

Corollary 3.5 ([26, Corollary 1.7]). *A projective module M is semiregular if and only if M/N has a projective cover for every finitely generated submodule N of M.*

Theorem 3.6 ([26, Theorem 1.12]). *Let M be a countably generated semiregular module and suppose that $\mathrm{Rad}(M)$ is small. Then $M \cong \bigoplus_{i=1}^{\infty} e_i R$ where $e_i^2 = e_i \in R$. In particular, M is projective.*

Corollary 3.7 ([26, Corollary 1.18]). *A finitely generated projective module M is semiregular if and only if it satisfies the following conditions.*

1) *Every finitely generated submodule of $M/\mathrm{Rad}(M)$ is a direct summand.*

2) *Direct decompositions of $M/\mathrm{Rad}(M)$ can be lifted to M.*

4 Semiregularity for Hom

We now consider $H := \mathrm{Hom}_R(A, M)$. We saw in Chapter IV, "Regularity in Modules" that H can naturally be considered a module over six possibly different rings, in particular the rings $S := \mathrm{End}(M_R)$ and $T := \mathrm{End}(A_R)$. For each of these there is the concept of semiregular according to Nicholson. We will consider another natural definition.

For the definition of $\mathrm{Rad}(A, M)$ see Lemma I.2.2.

Definition 4.1. The map $f \in \mathrm{Hom}_R(A, M)$ is **semiregular** or **S-semiregular** or **T-semiregular** if there exists a regular or S-regular or T-regular map $h \in H$, respectively, such that

$$h - f \in \mathrm{Rad}(A, M). \tag{9}$$

This definition differs from Nicholson's because $\mathrm{Rad}(A, M)$, $\mathrm{Rad}(_S H)$, and $\mathrm{Rad}(H_T)$ may differ. However, it is known ([20, Corollary 2.3]) that

$$\mathrm{Rad}(_S H) \cup \mathrm{Rad}(H_T) \subseteq \mathrm{Rad}(A, M).$$

This means that there are potentially more semiregular, S-, and T-semiregular maps according to Definition 4.1 than there are according to Nicholson's definition. On the other hand, we have also seen (Lemma 2.1) that $f \in H$ regular is stronger than either S-regular or T-regular.

We verify a remark for a semiregular map f and give a characterization of an S-regular map f. We work with S-regularity which is forced more or less by the fact that we have to apply the dual basis lemma for projective modules.

Remark. Let $f \in \mathrm{Hom}_R(A, M)$ and assume that f is semiregular. Then

1) f is $\mathrm{Rad}(A, M)$-regular;

2) f is partially invertible.

Proof. 1) By assumption we have $h - f = u \in \mathrm{Rad}(A, M)$ where h is regular. Then $h = f + u$ and there exists $g \in \mathrm{Hom}_R(M, A)$ such that $h = hgh$ and as $h = f + u$,

$$f + u = (f + u)g(f + u) = fgf + ug(f + u) + (f + u)gu.$$

This implies that

$$f - fgf = -u + ug(f + u) + (f + u)gu \in \mathrm{Rad}(A, M),$$

and hence f is $\mathrm{Rad}(A, M)$-regular.

2) Again we have a regular map $h = f + u$ where $u \in \mathrm{Rad}(A, M)$. By way of contradiction assume that f is not partially invertible. Then $f \in \mathrm{Tot}(A, M)$ and

$$h = f + u \in \mathrm{Tot}(A, M) + \mathrm{Rad}(A, M) = \mathrm{Tot}(A, M) \tag{10}$$

by [20, Theorem 2.4]. However, $h = hgh$ for some $g \in \mathrm{Hom}_R(M, A)$ and $e := hg = e^2$ hence h is partially invertible contrary to (10). $\square$

We now prove an interesting theorem that connects semiregularity with other standard notions.

Theorem 4.2. *If $f \in \mathrm{Hom}_R(A, M)$ is semiregular, then Sf lies over an S-projective direct summand of $_S \mathrm{Hom}_R(A, M)$.*

Although f semiregular implies that f is S-semiregular, Nicholson's result cannot be applied here because $\mathrm{Rad}(A, M)$ is not known to be contained in $\mathrm{Rad}(_S \mathrm{Hom}_R(A, M))$ and therefore f is not known to be semiregular in $_S \mathrm{Hom}_R(A, M)$ in the sense of Nicholson.

Proof. By assumption we have $h - f \in \mathrm{Rad}(A, M)$ with $hgh = h$ for some $g \in \mathrm{Hom}_R(M, A)$. Since $h - f \in \mathrm{Rad}(A, M)$ it follows that $(h - f)g \in \mathrm{Rad}(S)$. Then there exists $s_0 \in S$ such that $(1 - (h - f)g)s_0 = 1$. This implies that

$$fgs_0 = 1 - (1 - hg)s_0.$$

Multiplying by hg on the left and using that $(hg)^2 = hg$, we get

$$hgfgs_0 = hg.$$

Now multiplying by hgf on the right, we get

$$hgfgs_0hgf = hghgf = hgf,$$

and this shows that hgf is regular. Hence hgf has all the good properties of regular homomorphisms. In particular, $Shgf$ is a projective direct summand of $_SH$. Since hg is an idempotent in S, we have that $Shgf \subseteq Sf$. This is the first of the conditions that we had to satisfy.

Now we set

$$h_1 := hgf, \quad g_1 := gs_0, \quad e_1 := g_1h_1 \in T.$$

It follows that

$$h_1e_1 = h_1g_1h_1 = h_1 \tag{11}$$

and further

$$Sh_1 = Sh_1e_1 \subseteq He_1 = Hg_1h_1 \subseteq Sh_1,$$

hence

$$He_1 = Sh_1.$$

It now follows with the idempotent e_1 that

$$_SH = He_1 \oplus H(1 - e_1) = Sh_1 \oplus H(1 - e_1).$$

We still have to show that

$$U := Sf \cap _SH(1 - e_1) \subseteq \operatorname{Rad}(A, M).$$

We first observe that $U = Lf$ for some left ideal L of S. Then we show that $f - h_1 =: v \in \operatorname{Rad}(A, M)$. To do so, write $f = h - u$ with $u \in \operatorname{Rad}(A, M)$. Then

$$
\begin{aligned}
v = f - h_1 &= f - hgf = h - u - hg(h - u) \\
&= h - hgh - u + hgu = -u + hgu \in \operatorname{Rad}(A, M).
\end{aligned}
$$

Since $h_1(1 - e_1) = 0$ by (11),

$$
\begin{aligned}
U = U(1 - e_1) &= Lf(1 - e_1) = L(h_1 + v)(1 - e_1) \\
&= Lv(1 - e_1) \subseteq \operatorname{Rad}(A, M),
\end{aligned}
$$

and the theorem is proved.   □

Theorem 4.2 raises the immediate question whether the converse is true. We can use the results of Nicholson to show that f is S-semiregular, but it is open whether this also means that f is semiregular as a map.

Theorem 4.3. *Let* $f \in \mathrm{Hom}_R(A, M)$. *If* Sf *lies over an* S-*projective* S-*direct summand of* $_S \mathrm{Hom}_R(A, M)$, *then* f *is* S-*semiregular.*

Proof. By Proposition 3.3.4, f is semiregular as a element of the module $_SH$, hence there is a map $g \in H$ that is regular as an element of the module $_SH$, i.e., S-regular, and $f - g \in \mathrm{Rad}(_SH)$. Since $\mathrm{Rad}(_SH) \subseteq \mathrm{Rad}(A, M)$, the map f is S-semiregular in the sense of Definition 4.1. □

Question 4.4. If $f \in \mathrm{Rad}(A, M)$, then by definition of $\mathrm{Rad}(A, M)$ for every $g \in \mathrm{Hom}_R(A, M)$ it follows that $fg \in \mathrm{Rad}(S)$ and, equivalently, $gf \in \mathrm{Rad}(T)$. If $\sigma \in \mathrm{Hom}_S(_SH, S)$ and $f \in \mathrm{Rad}(A, M)$, is $f\sigma \in \mathrm{Rad}(S)$? We conjecture that this is true.

Chapter VII

$\mathrm{Reg}(A, M)$ and Other Substructures of Hom

1 Substructures of Hom

We turn to connections between $\mathrm{Reg}(A, M)$ and other substructures of $H :=$ $\mathrm{Hom}_R(A, M)$.

By $X \subseteq^* A$ we mean that X is large (= essential) in A, and by $X \subseteq^\circ A$ we mean that X is small (= superfluous) in A. Interesting substructures of Hom are the following.

The **singular submodule** of $_S H_T$ is by definition the S-T-submodule

$$\Delta(A, M) := \{f \in \mathrm{Hom}_R(A, M) \mid \mathrm{Ker}(f) \subseteq^* A\}.$$

The **cosingular submodule** of H is the S-T-submodule

$$\nabla(A, M) := \{f \in \mathrm{Hom}_R(A, M) \mid \mathrm{Im}(f) \subseteq^\circ M\}.$$

We also call $f \in \mathrm{Hom}_R(A, M)$ **singular** if $\mathrm{Ker}(f) \subseteq^* A$, and **cosingular** if $\mathrm{Im}(f) \subseteq^\circ M$.

$\Delta(A, M)$ and $\nabla(A, M)$ are not only S-T-bimodules but "ideals" in Mod-R having the following properties.

Lemma 1.1.

1) *If $f_1, f_2 \in \Delta(A, M)$, then $f_1 + f_2 \in \Delta(A, M)$.*

2) *If $X, Y \in$ Mod-R, $f \in \Delta(A, M)$, $g \in \mathrm{Hom}_R(M, Y)$, and $h \in \mathrm{Hom}_R(X, A)$, then $gfh \in \Delta(X, Y)$.*

3) *If $f_1, f_2 \in \nabla(A, M)$, then $f_1 + f_2 \in \nabla(A, M)$.*

    4) *If $X, Y \in$ Mod-R, $f \in \nabla(A, M)$, $g \in \mathrm{Hom}_R(M, Y)$, and $h \in \mathrm{Hom}_R(X, A)$, then $gfh \in \nabla(X, Y)$.*

Proof. 1) $\mathrm{Ker}(f_1) \cap \mathrm{Ker}(f_2) \subseteq \mathrm{Ker}(f_1 + f_2)$. As the intersection of two large submodules is large, also $\mathrm{Ker}(f_1) \cap \mathrm{Ker}(f_2)$ is large and so $\mathrm{Ker}(f_1 + f_2)$ is large. Hence $f_1 + f_2 \in \Delta(A, M)$.

    2) Suppose that $f \in \Delta(A, M)$, i.e., $f \in \mathrm{Hom}_R(A, M)$ and $\mathrm{Ker}(f)$ is large. Let $g \in \mathrm{Hom}_R(M, Y)$. As $\mathrm{Ker}(f) \subseteq \mathrm{Ker}(gf)$ it follows that $\mathrm{Ker}(gf)$ is large and $gf \in \Delta(A, Y)$. Let $h \in \mathrm{Hom}_R(X, A)$. To show that $\mathrm{Ker}(fh)$ is large, let $0 \neq B \subseteq X$ be given. In case $B \subseteq \mathrm{Ker}(h)$ we have $B \subseteq \mathrm{Ker}(fh)$ which shows that $\mathrm{Ker}(fh)$ intersects B non-trivially. In case $B \not\subseteq \mathrm{Ker}(h)$, we have that $h(B) \neq 0$. Since $\mathrm{Ker}(f) \subseteq^* A$, it follows that $\mathrm{Ker}(f) \cap h(B) \neq 0$. Let $b \in B$ with $0 \neq h(b) \in \mathrm{Ker}(f)$. Then $fh(b) = 0$, hence $B \cap \mathrm{Ker}(fh) \neq 0$. Together we have that $\mathrm{Ker}(fh) \subseteq^* X$ in all cases.

    3) $\mathrm{Im}(f_1 + f_2) \subseteq \mathrm{Im}(f_1) + \mathrm{Im}(f_2)$ and the sum of two small submodules is clearly small. So $\mathrm{Im}(f_1 + f_2)$ is also small and $f_1 + f_2 \in \nabla(A, M)$.

    4) Suppose that $f \in \nabla(A, M)$, i.e., $f \in \mathrm{Hom}_R(A, M)$ and $\mathrm{Im}(f)$ is small. Let $h \in \mathrm{Hom}_R(X, A)$. Then $\mathrm{Im}(fh) \subseteq \mathrm{Im}(f) \subseteq^\circ M$ so $\mathrm{Im}(fh) \subseteq^\circ M$. Let $g \in \mathrm{Hom}_R(M, Y)$ and suppose that $Y_1 + \mathrm{Im}(gfh) = Y$. We will show that $Y_1 = Y$. Let $x \in M$, then $g(x) = y_1 + gfh(x_1)$ for some $y_1 \in Y_1$ and some $x_1 \in X$. Then $y_1 = g(x - fh(x_1))$, so $x = (x - fh(x_1)) + fh(x_1) \in g^{-1}(Y_1) + \mathrm{Im}(fh)$. As $\mathrm{Im}(fh)$ is small in M, we have $g^{-1}(Y_1) = M$, so $Y_1 \subseteq g(M)$ and therefore $Y = Y_1 + \mathrm{Im}(gfh) = Y_1$. $\qquad\square$

    There are three different concepts of radical in H. There are the radicals of H as left S-module, and of H as right T-module, namely

$$\mathrm{Rad}(_SH) := \{f \in H \mid Sf \subseteq^\circ {}_SH\}, \ \mathrm{Rad}(H_T) := \{f \in H \mid fT \subseteq^\circ H_T\}.$$

The most important radical is the following for which there are two equivalent definitions (Lemma Ion radicals).

$$
\begin{aligned}
\mathrm{Rad}(H) \ &= \ \mathrm{Rad}(\mathrm{Hom}_R(A, M)) = \mathrm{Rad}(A, M) \\
&:= \ \{f \in H \mid f\,\mathrm{Hom}_R(M, A) \subseteq \mathrm{Rad}(S)\} \qquad (1) \\
&= \ \{f \in H \mid \mathrm{Hom}_R(M, A)f \subseteq \mathrm{Rad}(T)\}.
\end{aligned}
$$

It is easy to see that

$$\mathrm{Rad}(_SH), \mathrm{Rad}(H_T) \subseteq \mathrm{Rad}(H),$$

and that for any A, M,

$$\Delta(A, M), \nabla(A, M), \mathrm{Rad}(A, M) \subseteq \mathrm{Tot}(A, M).$$

A natural question arises: For which pairs A, M does the total equal one or more of the other structures?

    The sub-bimodules $\Delta(A, M)$, $\nabla(A, M)$, and $\mathrm{Rad}(A, M)$ all have intersection $\{0\}$ with $\mathrm{Reg}(A, M)$.

Proposition 1.2.

1) $\text{Reg}(A, M) \cap \Delta(A, M) = \{0\}$.

2) $\text{Reg}(A, M) \cap \nabla(A, M) = \{0\}$.

3) $\text{Reg}(A, M) \cap \text{Rad}(A, M) = \{0\}$

Proof. 1) Indeed, if $0 \neq f \in \text{Reg}(A, M)$, then $\text{Ker}(f) \neq A$ and $\text{Ker}(f) \subseteq^{\oplus} A$, so $\text{Ker}(f)$ is not large in A and $f \notin \Delta(A, M)$.

2) If $0 \neq f$ is regular, then $0 \neq \text{Im}(f) \subseteq^{\oplus} M$, hence $\text{Im}(f)$ is not small in M and $f \notin \nabla(A, M)$.

3) If $0 \neq f$ is regular with $fgf = f$, then $e := fg = e^2 \neq 0$ and e cannot be in $\text{Rad}(S)$. $\square$

$\text{Reg}(A, M)$ and $\text{Tot}(A, M)$ are on opposite substructures.

Proposition 1.3.

1) $\text{Reg}(A, M) \cap \text{Tot}(A, M) = 0$.

2) $\text{Tot}(S) \, \text{Reg}(A, M) = \text{Reg}(A, M) \, \text{Tot}(T) = 0$.

Proof. 1) This follows because every nonzero regular homomorphism is partially invertible.

2) Let $s \in \text{Tot}(S)$, and $f \in \text{Reg}(A, M)$. Since $\text{Reg}(A, M)$ is a left S-module, we have $sf \in \text{Reg}(A, M)$. Assume $sf \neq 0$. Then since sf is regular, it is partially invertible. Then by Lemma 2.3.1, s must be partially invertible, a contradiction. $\square$

Question 1.4. For which pairs A, M is

$$\text{Hom}_R(A, M) = \text{Reg}(A, M) \oplus \text{Tot}(A, M)?$$

What can be said about $\Sigma \, \text{Tot}(A, M)$, the additive closure of $\text{Tot}(A, M)$?

As $\text{Rad}(A, M) \subseteq \text{Tot}(A, M)$ ([20, Theorem 2.4]) the new Proposition 1.3.1 generalizes Proposition 1.2.3.

Example 1.5. Suppose that A and M are such that $\text{Hom}_R(M, A) = 0$. Then it is easily seen that

- $\text{Rad}(\text{Hom}_R(A, M) = \text{Hom}_R(A, M)$;

- $\text{Reg}(A, M) = 0$;

- $\text{Tot}(A, M) = \text{Hom}_R(A, M)$.

We provide some simple examples from abelian group theory.

Example 1.6. Let $A = \mathbb{Z} \oplus \mathbb{Z}$ and $M = \mathbb{Q}$. Then $\text{Hom}(M, A) = 0$ and $\text{Hom}(A, M) \cong \mathbb{Q} \oplus \mathbb{Q}$.

1) For any $f \in \mathrm{Hom}(A, M)$, we have $A/\mathrm{Ker}(f) \cong \mathrm{Im}(f) \subseteq \mathbb{Q}$. So $\mathrm{Ker}(f)$ is not large in A (Proposition I.3.8) unless $f = 0$. We conclude that $\Delta(A, M) = 0$.

2) For any $f \in \mathrm{Hom}(A, M)$, the image $\mathrm{Im}(f)$ is finitely generated and therefore has no divisible epimorphic image. Applying Proposition I.3.7 we see that $\mathrm{Im}(f)$ is small in $M = \mathbb{Q}$ and we conclude that $\nabla(A, M) = \mathrm{Hom}(A, M)$.

Example 1.7. Let $A = \mathbb{Z}(p^2)$ and $M = \mathbb{Z}(p^\infty)$. Then $\mathrm{Hom}(M, A) = 0$ and $\mathrm{Hom}(A, M) \cong \mathbb{Z}(p^2)$.

1) For any $f \in \mathrm{Hom}(A, M)$, we have that $\mathrm{Ker}(f)$ is 0 or $p\mathbb{Z}(p^2)$ or $\mathbb{Z}(p^2)$. This means that the kernel is large if and only if $p\mathbb{Z}(p^2) \subseteq \mathrm{Ker}(f)$. So the homomorphisms with large kernel are exactly those that factor through $\mathbb{Z}(p^2)/p\mathbb{Z}(p^2)$. We conclude that $\Delta(A, M) \cong \mathrm{Hom}(\mathbb{Z}(p^2)/p\mathbb{Z}(p^2), \mathbb{Z}(p^\infty)) \cong \mathbb{Z}(p)$.

2) For any $f \in \mathrm{Hom}(A, M)$, the image $\mathrm{Im}(f) \subseteq \mathbb{Z}(p^\infty)$ is always small (Proposition I.3.7) and we conclude that $\nabla(A, M) = \mathrm{Hom}(A, M)$.

Example 1.8. Let $A = \mathbb{Z}$ and $M = \mathbb{Z}(p)$. Then $\mathrm{Hom}(M, A) = 0$ and $\mathrm{Hom}(A, M) \cong \mathbb{Z}(p)$.

1) For any $f \in \mathrm{Hom}(A, M)$, we have that $\mathrm{Ker}(f)$ is large. We conclude that $\Delta(A, M) = \mathrm{Hom}(A, M)$.

2) For any $f \in \mathrm{Hom}(A, M)$, the image $\mathrm{Im}(f) \subseteq \mathbb{Z}(p)$ is not small unless $\mathrm{Im}(f) = 0$. We conclude that $\nabla(A, M) = 0$.

Example 1.9. Let p and q be different prime numbers, let $A = \mathbb{Z}(p) \oplus \mathbb{Z}(q^2)$ and $M = A$. Then

$$\mathrm{Hom}(A, M) \cong \mathrm{Hom}(\mathbb{Z}(p), \mathbb{Z}(p)) \oplus \mathrm{Hom}(\mathbb{Z}(q^2), \mathbb{Z}(q^2)) \cong \mathbb{Z}(p) \oplus \mathbb{Z}(q^2),$$

while

$$\mathrm{Reg}(A, M) \cong \mathrm{Hom}(\mathbb{Z}(p), \mathbb{Z}(p)) \cong \mathbb{Z}(p),$$

so $0 \neq \mathrm{Reg}(A, M) \neq \mathrm{Hom}(A, M)$.

Proof. Because of orders $\mathrm{Hom}(\mathbb{Z}(p), \mathbb{Z}(q^2)) = 0$ and $\mathrm{Hom}(\mathbb{Z}(q^2), \mathbb{Z}(p)) = 0$. Since the cyclic group $\mathbb{Z}(p)$ of order p is simple, $\mathrm{Reg}(\mathbb{Z}(p), \mathbb{Z}(p)) = \mathrm{Hom}(\mathbb{Z}(p), \mathbb{Z}(p))$ while $\mathrm{Reg}(\mathbb{Z}(q^2), \mathbb{Z}(q^2)) = 0$ because $\mathbb{Z}(q^2)$ is indecomposable but its endomorphism ring is not a division ring. ☐

Example 1.10. Let $A = \bigoplus_{p \in \mathbb{P}} \mathbb{Z}(p)$ and $M = \prod_{p \in \mathbb{P}} \mathbb{Z}(p)$. Then A is equal to the maximal torsion subgroup of M, and

$$
\begin{aligned}
\mathrm{Hom}(A, M) &= \mathrm{Hom}(A, A) = \mathrm{Hom}\left(\bigoplus_{p \in \mathbb{P}} \mathbb{Z}(p), \bigoplus_{p \in \mathbb{P}} \mathbb{Z}(p)\right) \\
&\cong \prod_{p \in \mathbb{P}} \mathrm{Hom}\left(\mathbb{Z}(p), \bigoplus_{p \in \mathbb{P}} \mathbb{Z}(p)\right) \cong \prod_{p \in \mathbb{P}} \mathrm{Hom}(\mathbb{Z}(p), \mathbb{Z}(p)),
\end{aligned}
$$

while
$$\text{Reg}(A, M) = \bigoplus_{p \in \mathbb{P}} \text{Hom}(\mathbb{Z}(p), \mathbb{Z}(p)).$$

Again $0 \neq \text{Reg}(A, M) \neq \text{Hom}(A, M)$.

Proof. Note that A is elementary, i.e., semisimple,

$$\text{End}(A) \cong \prod_{p \in \mathbb{P}} \text{Hom}(\mathbb{Z}(p), \bigoplus_{p \in \mathbb{P}} \mathbb{Z}(p)) = \prod_{p \in \mathbb{P}} \text{End}(\mathbb{Z}(p)),$$

and similarly,

$$\text{End}(M) \cong \prod_{p \in \mathbb{P}} \text{End}(\mathbb{Z}(p)).$$

A map $f \in \text{Hom}(A, M)$ is regular if and only if $\text{Im}(f)$ is a direct summand of M and this is the case if and only if $\text{Im}(f)$ is finite. The reason for this is that for any $f \in \text{Hom}(A, M) = \text{Hom}(A, A)$, we have $\text{Im}(f) = \bigoplus_{p \in P} \mathbb{Z}(p)$ for some set P of primes and $\prod_{p \in P} \mathbb{Z}(p) / \bigoplus_{p \in P} \mathbb{Z}(p)$ is divisible and therefore $\bigoplus_{p \in P} \mathbb{Z}(p)$ cannot be a direct summand of $\prod_{p \in P} \mathbb{Z}(p)$ and hence not of $\prod_{p \in \mathbb{P}} \mathbb{Z}(p)$ unless $\prod_{p \in P} \mathbb{Z}(p) / \bigoplus_{p \in P} \mathbb{Z}(p) = 0$, which is equivalent to P being finite. The maps in $\text{Hom}(A, M) = \prod_{p \in \mathbb{P}} \mathbb{Z}(p)$ with finite image are exactly those in $\bigoplus_{p \in \mathbb{P}} \mathbb{Z}(p)$. This subgroup of regular elements is evidently an S-T-submodule hence is the largest regular bi-submodule of $\text{Hom}(A, M)$. $\qquad\square$

2 Properties of $\Delta(A, M)$ and $\nabla(A, M)$

Let R be a ring and denote by $\text{Rad}(R)$ the radical of R. The radical has the following well-known property: For a left or right ideal A of R it is true that

$$A \subseteq \text{Rad}(R) \Leftrightarrow \forall a \in A, \ 1 - a \text{ is an invertible element of } R. \qquad (2)$$

Now consider $H := \text{Hom}_R(A, M)$, $S := \text{End}(M_R)$, $T := \text{End}(A_R)$. Then H is an S-T-bimodule. For $f \in H$ and $g \in \text{Hom}_R(M, A)$ we have $gf \in T$, $fg \in S$. Thus $\text{Hom}_R(M, A)f$ is a left ideal in T and $f\text{Hom}_R(M, A)$ is a right ideal in S.

We ask whether (2) can be applied with gf or fg in place of a but with weaker or different conditions replacing invertibility.

We need further notions (see [20]).

Definition 2.1.

1) A modules V is called **locally injective** if and only if for every submodule $A \subseteq V$ that is not large (= essential) in V, there exists a nonzero injective submodule Q of V with $A \cap Q = 0$.

2) A modules W is called **locally projective** if and only if for every submodule $B \subseteq W$ that is not small in W, there exists a nonzero projective direct summand P of W with $P \subseteq B$.

3) A module N is called **large restricted** if and only if every monomorphism $f : N \to N$ with $\text{Im}(f) \subseteq^* N$ is an automorphism, i.e., $\text{Im}(f) = N$.

4) A module N is called **small restricted** if and only if every epimorphism $f : N \to N$ with $\mathrm{Ker}(f) \subseteq^\circ N$ is an automorphism, i.e., $\mathrm{Ker}(f) = 0$.

The following results are easily verified (see [20]).

- *Every injective module is locally injective.*

- *Every projective and semiperfect module is locally projective.*

- *Every injective module is large restricted.*

- *Every projective module is small restricted.*

Remark. ([20]) If R_R is not Noetherian, then there exist infinite direct sums of injective R-modules that are not injective. But these are locally injective! We would like to have an example of a ring R such that R_R is locally injective but not injective. Also, we know no example of a locally projective ring that is not projective.

Theorem 2.2 ([20]).

1) *The module A is locally injective if and only if for all $M \in \mathrm{Mod}\text{-}R$, $\Delta(A, M) = \mathrm{Tot}(A, M)$.*

2) *The module M is locally projective if and only if for all $A \in \mathrm{Mod}\text{-}R$, $\nabla(A, M) = \mathrm{Tot}(A, M)$.*

3) *If A is large restricted, then $\Delta(A, M) \subseteq \mathrm{Rad}(A, M)$ for all $M \in \mathrm{Mod}\text{-}R$.*

4) *If M is small restricted, then $\nabla(A, M) \subseteq \mathrm{Rad}(A, M)$ for all $A \in \mathrm{Mod}\text{-}R$.*

Combining these properties we obtain the following corollary.

Corollary 2.3 ([20]).

1) *If A is locally injective and M is locally projective, then*

$$\Delta(A, M) = \nabla(A, M) = \mathrm{Tot}(A, M).$$

2) *If A is injective and M is locally projective or if A is locally injective and M is projective and semi-perfect, then*

$$\Delta(A, M) = \nabla(A, M) = \mathrm{Rad}(A, M) = \mathrm{Tot}(A, M).$$

We include a variant of such results with proof.

Proposition 2.4. *If A or M is large restricted or injective, then*

$$\Delta(A, M) \subseteq \mathrm{Rad}(A, M).$$

Proof. Let $f \in \Delta(A, M)$, i.e., $\mathrm{Ker}(f) \subseteq^* A$ and let $g \in \mathrm{Hom}_R(M, A)$. Then also $\mathrm{Ker}(gf) \subseteq^* A$ since $\mathrm{Ker}(f) \subseteq \mathrm{Ker}(gf)$. But also $\mathrm{Ker}(fg) \subseteq^* M$. To see this, let $0 \neq U \subseteq M$. If $g(U) = 0$, then $U \subseteq \mathrm{Ker}(fg)$ and all is well. So suppose that $g(U) \neq 0$. Then $\mathrm{Ker}(f) \cap g(U) \neq 0$ since $\mathrm{Ker}(f) \subseteq^* A$. Hence there exists $0 \neq u \in U$ such that $g(u) \neq 0$ but $fg(u) = 0$. This means that $\mathrm{Ker}(fg) \subseteq^* M$.

Consider
$$a \in \mathrm{Ker}(1 - gf) \cap \mathrm{Ker}(gf).$$
Then
$$0 = (1 - gf)a = a - gf(a) = a,$$
hence $\mathrm{Ker}(1 - gf) \cap \mathrm{Ker}(gf) = 0$. Since $\mathrm{Ker}(gf) \subseteq^* A$, this implies $\mathrm{Ker}(1 - gf) = 0$, hence $1 - gf$ is a monomorphism. It follows similarly that $1 - fg$ is a monomorphism. Further we have
$$\mathrm{Ker}(gf) \subseteq \mathrm{Im}(1 - gf)$$
which implies that $\mathrm{Im}(1 - gf) \subseteq^* A$. If A is large restricted, then it follows that $1 - gf$ is also an epimorphism, so altogether an automorphism. If A is injective, then it is large restricted. It follows similarly that $1 - fg$ is an automorphism if M is large restricted, and in particular if M is injective. Now apply (2) and consider the definition of $\mathrm{Rad}(A, M)$. $\square$

This is again a result connecting two substructures of $\mathrm{Hom}(A, M)$.

Next we consider $\nabla(A, M)$. Here for $f \in \nabla(A, M)$, $\mathrm{Im}(f) \subseteq^\circ M$ and for $g \in \mathrm{Hom}_R(M, A)$, it follows that also $\mathrm{Im}(gf) \subseteq^\circ A$, $\mathrm{Im}(fg) \subseteq^\circ M$. Then from
$$\mathrm{Im}(1 - gf) + \mathrm{Im}(gf) = A, \quad \mathrm{Im}(1 - fg) + \mathrm{Im}(fg) = M$$
it follows that $\mathrm{Im}(1 - gf) = A$, $\mathrm{Im}(1 - fg) = M$, hence $1 - gf$, $1 - fg$ are epimorphisms.

Proposition 2.5. *Let $f \in \nabla(A, M)$.*

1) *If A is projective, then for all $g \in \mathrm{Hom}_R(M, A)$, the map $1 - gf$ is regular.*

2) *If M is projective, then for all $g \in \mathrm{Hom}_R(M, A)$ the map $1 - fg$ is regular.*

Proof. 1) We know that $1 - gf$ is an epimorphism on the projective module A, and therefore splits. Thus $\mathrm{Im}(1 - gf)$ and $\mathrm{Ker}(1 - gf)$ are direct summands which implies that $1 - gf$ is regular.

2) Same as 1). $\square$

3   The Special Case $\mathrm{Hom}_R(R, M)$

Until now the singular submodules did not occur explicitly. We show now that they are already included as a special case in our general considerations.

Finally we consider two special cases of this result; first the case $A = R$, M arbitrary, and then the case $A = M = R$. Here we apply the isomorphism

$$\rho : \operatorname{Hom}_R(R, M) \ni f \mapsto f(1) \in M$$

from $\operatorname{Hom}_R(R, M)$ to M, and from $\operatorname{Hom}_R(R, R)$ to R. Then

$$\Delta(M_R) := \rho(\Delta(R, M) = \{m \in M \mid \operatorname{Ann}_R(m) \subseteq^* R_R\},$$

that is, $\Delta(M_R)$ is properly the singular submodule of M. Here $S := \operatorname{End}(M_R)$ as before, but $T = \operatorname{End}(R_R) \cong R$.

Corollary 3.1.

1) *If R_R or M_R is injective, then $\Delta(M_R) \subseteq \operatorname{Rad}(M_R)$.*

2) *If R_R is injective, then $\Delta(R_R) \subseteq \operatorname{Rad}(R)$.*

We again consider the special case $A = R$ and translate using ρ. Then

$$f \in \nabla(R, M) \Leftrightarrow f(1)R \subseteq^\circ M \Leftrightarrow m := f(1) \in \operatorname{Rad}(M),$$

i.e., $\nabla(M_R) := \rho(\nabla(R, M)) = \operatorname{Rad}(M)$.

It follows for all $g \in \operatorname{Hom}_R(M, R)$ that $g(m) \in \operatorname{Rad}(R)$ and from this we get the result that $1 - g(m)$ is not only regular but invertible. For projective M, the map $1 - mg$ is again regular. If we assume finally that $A = M = R$, then $m \in \operatorname{Rad}(R)$ and the action of $g \in \operatorname{Hom}_R(R, R)$ is the left multiplication in R by an element of R. Now gm and mg are in $\operatorname{Rad}(R)$ and $1 - gm$ and $1 - mg$ are invertible.

Question 3.2. If R_R is injective, is it true that for all $M \in \operatorname{Mod-}R$ we have $\operatorname{Rad}(M) = \operatorname{Tot}(M)$?

Question 3.3. For $\rho : \operatorname{Hom}_R(R, M) \ni f \mapsto f(1) \in M$, does it follow that $\rho(\operatorname{Rad}(R, M)) = \operatorname{Rad}(M)$?

We found that

$$\rho(\nabla(R, M)) = \operatorname{Rad}(M).$$

What is now $\rho(\operatorname{Rad}(R, M))$? By definition, $f \in \operatorname{Rad}(R, M)$ if and only if for all $\varphi \in M^* := \operatorname{Hom}_R(M, R)$,

$$(\varphi f)(r) = \varphi(f(1)r) = \varphi(f(1))r \in \operatorname{Rad}(R),$$

hence $\varphi(f(1)) \in \operatorname{Rad}(R)$. This implies that

$$\rho(\operatorname{Rad}(R, M)) = \{m \in M \mid \forall \varphi \in M^*, \varphi(m) \in \operatorname{Rad}(R)\}.$$

Since for any homomorphism φ,

$$\varphi(\operatorname{Rad}(M)) \subseteq \operatorname{Rad}(R), \text{ it follows that } \operatorname{Rad}(M) \subseteq \rho(\operatorname{Rad}(R, M)).$$

Finally, we have to describe $\rho(\operatorname{Tot}(R, M))$. Recall that $\operatorname{Tot}(M_R) := \{m \in M \mid m$ is not partially invertible.$\}$.

Proposition 3.4. $\rho(\mathrm{Tot}(R, M)) = \mathrm{Tot}(M_R)$.

Proof. Recall that $f \in \mathrm{Hom}_R(R, M)$ is partially invertible if and only if there exists $\varphi \in M^*$ such that

$$\varphi(f(r)) = \varphi(f(1))r$$

is left multiplication in R by the idempotent $\varphi(f(1))$. Hence

$$\rho(\mathrm{Tot}(R, M)) = \{m \in M \mid \forall\, \varphi \in M^*, \varphi(m) \text{ is not an idempotent} \neq 0\}.$$

By Lemma IV.2.3, $f \in \mathrm{Hom}_R(R, M)$ is partially invertible if and only if $f(1)$ is partially invertible. Hence f is not partially invertible if and only if $f(1)$ is not partially invertible. This is the claim. $\qquad\square$

We now specialize the results of Theorem 2.2 and Corollary 2.3 to the case $\mathrm{Hom}_R(R, M)$ and translate them by means of ρ into results about M.

Corollary 3.5. *Let R be a ring.*

1) *The module R_R is locally injective if and only if*

$$\forall\, M \in \text{Mod-}R: \ \Delta(M) = \mathrm{Tot}(M).$$

2) *If the module R_R is injective, then*

$$\forall\, M \in \text{Mod-}R: \ \Delta(M) = \mathrm{Rad}(M) = \mathrm{Tot}(M).$$

3) *If R_R is injective and W is locally projective, then*

$$\Delta(W) = \mathrm{Rad}(W) = \mathrm{Tot}(W).$$

4) *If R_R is locally injective and W is locally projective, then*

$$\Delta(W) = \mathrm{Rad}(W) = \mathrm{Tot}(W).$$

5) *If W is locally projective, then*

$$\mathrm{Rad}(W) = \mathrm{Tot}(W).$$

6) *If R_R is large restricted, then*

$$\forall\, M \in \text{Mod-}R: \ \Delta(M) \subseteq \rho(\mathrm{Rad}(R, M)).$$

We now show with an example that the equalities that have been established under certain assumptions, are not always true.

Example 3.6. Let R be a ring that is not right-Kasch. This means that there exists a simple right R-module B for which there does not exist an isomorphic right ideal of R. Then the only homomorphism from B to R is the 0-mapping. Hence $\mathrm{Rad}(B) = 0$ but $\rho(\mathrm{Rad}(R, B)) = \Delta(B) = \mathrm{Tot}(B) = B$.

Proof. To prove $\Delta(B) = B$, let $0 \neq b \in B$. Then $bR = B$ and the mapping $R \ni r \mapsto br \in bR = B$ is an epimorphism, hence $R/\mathrm{Ann}_R(b) \cong B$. Therefore $C := \mathrm{Ann}_R(b)$ is a maximal right ideal of R. We will show that C is large in R_R. Assume that $0 \neq U \subseteq R_R$ and $C \cap U = 0$. Since C is maximal it follows that $C \oplus U = R$. But then $B \cong R/C \cong U$, a contradiction to the choice of B. Since $C := \mathrm{Ann}_R(b)$ is large, b is singular, hence $\Delta(B) = B$.

What is now $\mathrm{Tot}(B)$? Since $\{0\} = \mathrm{Hom}_R(B, R)$, no element in B is partially invertible, hence $\mathrm{Tot}(B) = B$. □

4　Further Internal Properties of $\Delta(M)$

The advantage of identifying $\Delta(M)$ with $\rho(\Delta(R, M))$ rests on the fact that the general results about $\Delta(A, M)$ can be applied to this special case. We now consider more internal properties.

A module M is **singular** if $\Delta(M) = M$. It is easy to find examples of singular modules, e.g., by employing the following remark.

Remark. If $A \subseteq^* M$, then M/A is singular.

Proof. Set $\overline{m} = m + A \in M/A$. We have to show: If $0 \neq U$ is a right ideal of R, then $\mathrm{Ann}_R(\overline{m}) \cap U \neq 0$. Since U is a right ideal of R, mU is a submodule of M.

Case $mU = 0$. Then $U \subseteq \mathrm{Ann}_R(m) \subseteq \mathrm{Ann}_R(\overline{m})$.

Case $mU \neq 0$. Choose $u \in U$ such that $mu \neq 0$. Then $muR \neq 0$. Since $muR \subseteq M$ and $A \subseteq^* M$, it follows that $A \cap muR \neq 0$. Choose $0 \neq a = mur_0 \in A \cap muR$. Then $0 \neq ur_0 \in \mathrm{Ann}_R(\overline{m}) \cap U$ as needed. □

There are situations in which the Remark is used in the following form: If M/A is not singular, the A is not large in M.

A module is **nonsingular** if its only singular element is 0. Note that in contrast "not singular" means that at least one element is not singular.

By Zorn's Lemma it follows immediately that every module contains a maximal nonsingular submodule.

Lemma 4.1.

1) If A and B with $A \cap B = 0$ are nonsingular submodules of a module M, then $A + B = A \oplus B$ is a nonsingular submodule of M.

2) If $A \subseteq^* M$ and A is nonsingular, then M is nonsingular.

3) If $A \subseteq^* M$ and A is singular, then M does not contain a nonsingular submodule different from 0.

4) *If C is a maximal nonsingular submodule of M, then*

$$\Delta(M) \cap C = 0, \quad \Delta(M) + C = \Delta(M) \oplus C \subseteq^* M,$$

and C is a complement of $\Delta(C)$ in M.

5) *Let C be a maximal nonsingular submodule of M. If A is a complement of C in M with $\Delta(M) \subseteq A$, then $\Delta(M) \subseteq^* A$ and $A + C = A \oplus C \subseteq^* M$.*

Proof. 1) Let $0 \neq a \in A$, $b \in B$, and assume that $(a + b)r = ar + br = 0$ where $r \in R$. Then $ar = -br \in A \cap B = 0$, hence $ar = 0$. This implies that $\text{Ann}_R(a+b) \subseteq \text{Ann}_R(a)$. Since by assumption $\text{Ann}_R(a)$ is not large, also $\text{Ann}_R(a+b)$ is not large and hence $a + b$ is not singular.

2) The assumption that $\Delta(M) \neq 0$ would imply that $A \cap \Delta(M) \neq 0$, and A would not be nonsingular contrary to hypothesis.

3) Assume by way of contradiction that B is a nonsingular submodule of M. Then $A \cap B \neq 0$, hence A contains elements that are non-singular contrary to hypothesis.

4) By definition of $\Delta(M)$ we have that $\Delta(M) \cap C = 0$. Suppose $0 \neq U \subseteq M$. Assume that $(\Delta(M) + C) \cap U = 0$. Then U cannot contain a nonzero singular element since this would also belong to $\Delta(M)$. Hence U is nonsingular. Since $C \cap U = 0$, it follows by 1) that $C + U$ is nonsingular contrary to the maximality of C. Therefore $(\Delta(M) + C) \cap U \neq 0$.

Now let B be a complement of $\Delta(M)$ with $C \subseteq B$. Then B cannot contain a nonzero singular element since such an element would also belong to $\Delta(M)$. Hence B is nonsingular and equal to the maximal nonsingular submodule C.

5) This statement is true in general for complements but in our case there is a shorter proof. Assume that $0 \neq U \subseteq A$ and $\Delta(M) \cap U = 0$. Then U is nonsingular. Since $C \cap A = 0$, also $C \cap U = 0$. By 1) $C + U$ is nonsingular which contradicts the maximality of C. By 4) it follows that $A + C \subseteq^* M$. $\square$

We would like to apply our results also to abelian groups, that is, $\mathbb{Z}$-modules, but it is not more work to do this not only for $\mathbb{Z}$-modules but for (right) uniform rings. A ring R is called (right) **uniform** if and only if every nonzero right ideal of R is large in R_R, or, what means the same thing, any two nonzero right ideals have nonzero intersection. It is obvious that any commutative ring without zero divisors is uniform. If now M is a module over a uniform ring R and if for some $m \in M$, $\text{Ann}_R(m) \neq 0$, then $\text{Ann}_R(m)$ is a large right ideal, hence m is singular. This means: If m is torsion, then m is singular. Hence the maximal torsion submodule of M is equal to $\Delta(M)$ and M is torsion–free means that M is nonsingular. If m is not torsion $=$ singular, then mR is nonsingular; this we will use in the following.

Lemma 4.2. *Let R_R be uniform. Then $\Delta(M)$ is a complement of any maximal nonsingular submodule C of M.*

Proof. Let C be a maximal nonsingular submodule of M and let A be a complement of C in M with $\Delta(M) \subseteq A$. Assume that $a \in A$, $a \notin \Delta(M)$. Then aR is nonsingular. But since $C \cap A = 0$, also $C \cap aR = 0$, hence by Lemma 4.1.1, $C + aR$ is nonsingular in contradiction to the maximality of C. Hence $\Delta(M) = A$. $\qquad\square$

We need a definition from [25]. A module M_R is **quasi-continuous** if every complement in M is a direct summand, and if A and B are direct summands of M and $A \cap B = 0$, then $A \oplus B$ is a direct summand of M.

Theorem 4.3.

1) *Let C be a maximal nonsingular submodule of M and A a complement of C in M containing $\Delta(M)$. If M is quasi-continuous, then*

$$A \oplus C = M.$$

2) *If M is quasi-continuous, R_R is uniform, and C is a maximal nonsingular submodule of M, then*
$$\Delta(M) \oplus C = M.$$

Proof. 1) Since A and C are complements in a quasi-continuous module, they are direct summands. Since also $A \cap C = 0$, it follows that $A \oplus C$ is a direct summand of M. But $A \oplus C \subseteq^* M$ by Lemma 4.1.5, so $A \oplus C = M$.

2) By Lemma 4.2 we know that $A = \Delta(M)$; hence $A \oplus C = M$ implies that $\Delta(M) \oplus C = M$. $\qquad\square$

We return to the general situation making no assumptions on R or M. By now we have seen many interesting relations between $\Delta(M)$, the radical and the total, and also between $\Delta(M)$ and maximal nonsingular submodules. Yet, it is still difficult to find, for a module M, explicit properties of $\Delta(M)$ and a complement C of $\Delta(M)$ in M. In the following we exhibit some results concerning this topic.

We will have to use the following characterization of the socle of a module (see [18, 9.1.1]):
$$\mathrm{Soc}(M) = \bigcap_{B \subseteq^* M} B. \qquad (3)$$

In general, $\mathrm{Soc}(M)$ is not large in M, but if M is Artinian, then $\mathrm{Soc}(M) \subseteq^* M$ since it is the intersection of finitely many large submodules. For a module M_R and a right ideal I of R we set $\mathrm{Ann}_M(I) := \{m \in M \mid mI = 0\}$.

Theorem 4.4.

1) *For any module M, $\Delta(M) \subseteq \mathrm{Ann}_M(\mathrm{Soc}(R_R))$.*

2) *If $\mathrm{Soc}(R_R)$ is large in R_R, then $\Delta(M) \subseteq^* \mathrm{Ann}_M(\mathrm{Soc}(R_R))$.*

3) *If $\mathrm{Soc}(R_R)$ is large in R_R and if $\mathrm{Soc}(R_R) = \mathrm{Soc}(_RR)$, then $\mathrm{Rad}(R) \subseteq \Delta(R_R)$.*

Proof. 1) If $m \in \Delta(M)$, then it follows by (3) that $m\,\mathrm{Soc}(R_R) = 0$, hence $m \in \mathrm{Ann}_M(\mathrm{Soc}(R_R))$.

2) If $m \in \mathrm{Ann}_M(\mathrm{Soc}(R_R))$ and if $\mathrm{Soc}(R_R)$ is large in M, then $m \in \Delta(M)$.

3) If U is a left R-module, then $\mathrm{Rad}(R)U \subseteq \mathrm{Rad}(U)$. If U is semisimple, then $\mathrm{Rad}(U) = 0$, hence $\mathrm{Rad}(R)\,\mathrm{Soc}(_R R) = 0$. It follows that $\mathrm{Rad}(R)\,\mathrm{Soc}(R_R) = 0$ hence $\mathrm{Rad}(R) \subseteq \Delta(R_R)$. $\square$

In Corollary 3.5.2 we have seen that if the ring R_R is injective, then

$$\Delta(M) = \mathrm{Rad}(M) = \mathrm{Tot}(M) \tag{4}$$

for all $M \in \mathrm{Mod}\text{-}R$. This includes, for R_R injective, the identities

$$\Delta(R_R) = \mathrm{Rad}(R) = \mathrm{Tot}(R).$$

In $\mathrm{Rad}(R)$ and $\mathrm{Tot}(R)$ we did not indicate whether the ring is considered a right or left R-module. This is because the concepts are independent of the side of the R-action. For $\mathrm{Rad}(R)$ this is well-known. For $\mathrm{Tot}(R)$ it follows from the fact that "partially invertible" is independent of side.

There is another situation where (4) is valid.

First we need the following lemma which requires another definition from [25]. A module M_R is **continuous** if every complement in M is a direct summand, and every submodule isomorphic to a direct summand is itself a direct summand.

Lemma 4.5. ([25, Proposition 3.5]) *If M_R is continuous and $S := \mathrm{End}(M_R)$, then*

$$\mathrm{Rad}(S) = \Delta(M,M) = \{f \in S \mid \mathrm{Ker}(f) \subseteq^* M\}.$$

Theorem 4.6. *If R_R is continuous, then*

$$\Delta(R_R) = \mathrm{Rad}(R) = \mathrm{Tot}(R).$$

Proof. We apply Lemma 4.5 with R_R in place of M_R. Then S is now the ring of left multiplications on R by elements of R which we identify with R. Then we obtain by Lemma 4.5 that $\Delta(R_R) = \mathrm{Rad}(R)$. It remains to show that $\mathrm{Tot}(R) \subseteq \Delta(R_R)$. Let $b \in R$ with $\mathrm{Ann}_R(b)$ not large in R_R. Then there exists a complement K of $\mathrm{Ann}_R(b)$. Since R_R is continuous, the submodule K is a direct summand of R_R. Since $\mathrm{Ann}_R(b) \cap K = 0$, the mapping

$$K \ni x \mapsto bx \in bK$$

is an isomorphism. Since R_R is continuous and K is a direct summand, also bK is a direct summand. By Theorem II.2.1.4 it follows that b is partially invertible. We have now shown: If b is not singular, then it is partially invertible. Hence $\mathrm{Tot}(R) \subseteq \Delta(R_R)$. $\square$

5   Non-singular Modules

We have seen that any module contains a maximal nonsingular submodule. But what does it look like? Can we describe at least some nonsingular submodule? We give some information in this direction and start with simple modules.

Lemma 5.1. *Let U be a simple module. Then the following statements hold.*

1) *U is nonsingular if and only if U is projective.*

2) *If $U \subseteq M$ and $U \not\subseteq \mathrm{Rad}(M)$, then $U \subseteq^{\oplus} M$.*

3) *If $U \subseteq M$, U is nonsingular, and if C is a maximal nonsingular submodule of M, then $U \subseteq C$. Consequently, M contains a unique largest nonsingular submodule.*

Proof. 1) If $0 \neq u \in U$, then $uR = U$ since U is simple. The homomorphism

$$g : R \ni r \mapsto ur \in U$$

is an isomorphism and $\mathrm{Ker}(g) = \mathrm{Ann}_R(u)$. Then $R/\mathrm{Ann}_R(u) \cong U$ and since U is simple, $\mathrm{Ann}_R(u)$ is a maximal right ideal of R. Since U is nonsingular, $\mathrm{Ann}_R(u)$ is not a large right ideal of R. Hence there exists $0 \neq B \subseteq R_R$ such that $B \cap \mathrm{Ann}_R(u) = 0$.

Only if: Since $\mathrm{Ann}_R(u)$ is maximal and $B \neq 0$ it follows that $R = B + \mathrm{Ann}_R(u) = B \oplus \mathrm{Ann}_R(u)$. Then B is projective and as $B \cong R/\mathrm{Ann}_R(u) \cong U$, so is U.

If: Since now U is projective, the epimorphism g splits:

$$R_R = uR \oplus \mathrm{Ann}_R(u), \quad uR \neq 0.$$

Since $uR \cap \mathrm{Ann}_R(u) = 0$, $\mathrm{Ann}_R(u)$ is not large in R_R, hence u is not singular.

2) By hypothesis $U \not\subseteq \mathrm{Rad}(M)$. Therefore U is not small in M. Hence there exists $M_0 \subsetneq M$ such that
$$M = M_0 + U.$$

Since $M_0 \neq M$ it follows that $U \not\subseteq M_0$ and since U is simple $M_0 \cap U = 0$, hence $M = M_0 \oplus U$.

3) If U is a simple nonsingular submodule of M, then either $U \subseteq C$ or $U \cap C = 0$. By Lemma 4.1.1 the second case would imply that $C \oplus U$ is nonsingular contradicting the maximality of C. $\qquad \square$

We denote by $\mathrm{Doc}(M)$ the sum of all simple submodules of M that are not contained in $\mathrm{Rad}(M)$. Then $\mathrm{Doc}(M)$ is a direct summand of $\mathrm{Soc}(M)$, the socle of M:

$$\mathrm{Soc}(M) = \mathrm{Doc}(M) \oplus (\mathrm{Soc}(M) \cap \mathrm{Rad}(M)).$$

Theorem 5.2. *If M is projective or locally projective, then $\mathrm{Doc}(M)$ is a nonsingular and projective submodule of M. If C is a maximal nonsingular submodule of M, then $\mathrm{Doc}(M) \subseteq C$.*

Proof. Let U be a simple submodule of M such that $U \not\subseteq \mathrm{Rad}(M)$. Then by Lemma 5.1.2 $U \subseteq^{\oplus} M$. If M is projective, then so is U. If M is locally projective, then, since U is not small in M, we know that U contains a nonzero projective direct summand: $P \subseteq^{\oplus} M$. Since U is simple $U = P$. In either case U is nonsingular by Lemma 5.1.1. By Lemma 4.1.1 also $\mathrm{Doc}(M)$ is nonsingular. This is clear if $\mathrm{Doc}(M)$ is the sum of finitely many simple summands. If $\mathrm{Doc}(M)$ is the sum of infinitely many summands, and some element of $\mathrm{Doc}(M)$ is singular, then this element is contained in a finite subsum that is nonsingular. Hence the singular element must be 0. That $\mathrm{Doc}(M)$ is projective is clear. It follows from Lemma 5.1.3 that $\mathrm{Doc}(M) \subseteq C$. $\square$

We mention two special cases of the theorem.

Corollary 5.3.

1) *If R_R is a ring with $1 \in R$, then $\mathrm{Doc}(R_R)$ is nonsingular and projective.*

2) *If M is a projective or locally projective module with $\mathrm{Rad}(M) = 0$, then $\mathrm{Soc}(M)$ is nonsingular and projective.*

6 A Correspondence Between Submodules of $\mathrm{Hom}_R(A, M)$ and Ideals of $\mathrm{End}(M_R)$

In (1) we defined the bi-submodule $\mathrm{Rad}(A, M)$ by using $\mathrm{Rad}(S)$. But the reverse is also possible. Using $\mathrm{Rad}(A, M)$ we can define a two-sided ideal in S contained in $\mathrm{Rad}(S)$. Let

$$\mathrm{Idl}(\mathrm{Rad}(A, M)) := \sum\nolimits_{g \in \mathrm{Hom}_R(M, A)} \mathrm{Rad}(A, M)g.$$

Then this is a two-sided ideal in S since $\mathrm{Rad}(A, M)$ is a left S-module and $\mathrm{Hom}_R(M, A)$ is a T-S-bimodule. That $\mathrm{Idl}(\mathrm{Rad}(A, M)) \subseteq \mathrm{Rad}(S)$ follows from (1).

We generalize these constructions to arrive at a correspondence between the lattice of two-sided ideals of S and S-T-submodules of $H = \mathrm{Hom}_R(A, M)$.

Definition 6.1. Let I be a two-sided ideal in S, and K an S-T-submodule of H. Let

$$\mathrm{Mdl}(I) := \{f \in H \mid f \, \mathrm{Hom}_R(M, A) \subseteq I\},$$

and

$$\mathrm{Idl}(K) := \sum\nolimits_{g \in \mathrm{Hom}_R(M, A)} Kg.$$

We will see that $\mathrm{Mdl}(I)$ is an S-T-submodule of H, and $\mathrm{Idl}(K)$ is a two-sided ideal of S and that the maps Mdl and Idl have properties that are reminiscent of a Galois connection.

Lemma 6.2. *In the above context the following statements are valid.*

1) $\mathrm{Mdl}(I)$ *is an S-T-submodule of* $\mathrm{Hom}_R(A, M)$ *and* $\mathrm{Idl}(K)$ *is a two-sided ideal of S;*

2) *if* $I_1 \subseteq I_2$*, then* $\mathrm{Mdl}(I_1) \subseteq \mathrm{Mdl}(I_2)$ *and if* $K_1 \subseteq K_2$*, then* $\mathrm{Idl}(K_1) \subseteq \mathrm{Idl}(K_2)$;

3) $I \, \mathrm{Hom}_R(A, M) \subseteq \mathrm{Mdl}(I)$, $\mathrm{Idl}(K) \, \mathrm{Hom}_R(A, M) \subseteq K$;

4) $\mathrm{Idl}(\mathrm{Mdl}(I)) \subseteq I$, $K \subseteq \mathrm{Mdl}(\mathrm{Idl}(K))$;

5) $\mathrm{Mdl}(\mathrm{Idl}(\mathrm{Mdl}(I))) = \mathrm{Mdl}(I)$, *and* $\mathrm{Idl}(\mathrm{Mdl}(\mathrm{Idl}(K))) = \mathrm{Idl}(K)$;

6) *for bi-submodules* K_1*,* K_2 *of* $\mathrm{Hom}_R(A, M)$*,*

$$\mathrm{Idl}(K_1 + K_2) = \mathrm{Idl}(K_1) + \mathrm{Idl}(K_2);$$

7) *for ideals* I_1*,* I_2 *of* $\mathrm{End}(M_R)$*,*

$$\mathrm{Mdl}(I_1 \cap I_2) = \mathrm{Mdl}(I_1) \cap \mathrm{Mdl}(I_2).$$

Proof. 1) If $f_1 g, f_2 g \in I$, then $(f_1 + f_2)g \in I$ and $\mathrm{Mdl}(I)$ is additively closed. As I is a left ideal in S, if $fg \in I$, then also $s(fg) = (sf)g \in I$ so $\mathrm{Mdl}(I)$ is a left S-module. Since $\mathrm{Hom}_R(M, A)$ is a T-S-module, it follows that $tg \in \mathrm{Hom}_R(M, A)$ for any $t \in T$ and any $g \in \mathrm{Hom}_R(M, A)$. If $fg \in I$ for all g, then for all t and all g it follows that $f(tg) = (ft)g \in I$, hence $\mathrm{Mdl}(I)$ is also a right T-module.

It is similarly easy to see that $\mathrm{Idl}(K)$ is a two-sided ideal in S.

2) The definitions of $\mathrm{Mdl}(I)$ and $\mathrm{Idl}(K)$ make it clear that these constructions preserve inclusions.

3) For $s \in I$, $f \in H$, and $g \in \mathrm{Hom}_R(M, A)$ we have $(sf)g = s(fg)$, and since $fg \in S$ and I is an ideal, it follows that $s(fg) \in I$, hence $sf \in \mathrm{Mdl}(I)$. This says that $IH \subseteq \mathrm{Mdl}(I)$. Now suppose that $\sum_{i=1}^n h_i g_i \in \mathrm{Idl}(K)$ where $h_i \in K$, $g_i \in \mathrm{Hom}_R(M, A)$, and let $f \in H$. Then $(h_i g_i)f = h_i(g_i f)$. Since $g_i f \in T$ and K is a right T-module, it follows that $h_i(g_i f) \in K$, hence $\left(\sum_{i=1}^n h_i g_i \right) f = \sum_{i=1}^n h_i(g_i f) \in K$.

4) Here we need to deal with

$$\mathrm{Idl}(\mathrm{Mdl}(I)) = \sum_{g \in \mathrm{Hom}_R(M, A)} \mathrm{Mdl}(I)g.$$

If $f \in \mathrm{Mdl}(I)$, then $fg \in I$, hence $\mathrm{Mdl}(I)g \subseteq I$, and so $\mathrm{Idl}(\mathrm{Mdl}(I)) \subseteq I$. For the second inclusion, let $h \in K$, and $g \in \mathrm{Hom}_R(M, A)$. Then $hg \in \mathrm{Idl}(K)$ by definition of $\mathrm{Idl}(K)$, hence $h \in \mathrm{Mdl}(\mathrm{Idl}(K))$.

5) We use 4) and 2). From $\mathrm{Idl}(\mathrm{Mdl}(I)) \subseteq I$ it follows that

$$\mathrm{Mdl}(\mathrm{Idl}(\mathrm{Mdl}(I))) \subseteq \mathrm{Mdl}(I). \tag{5}$$

In $K \subseteq \mathrm{Mdl}(\mathrm{Idl}(K))$ we take $K = \mathrm{Mdl}(I)$ to obtain

$$\mathrm{Mdl}(I) \subseteq \mathrm{Mdl}(\mathrm{Idl}(\mathrm{Mdl}(I))).$$

This together with (5) establishes the first claim.
From $K \subseteq \mathrm{Mdl}(\mathrm{Idl}(K))$ it follows that

$$\mathrm{Idl}(K) \subseteq \mathrm{Idl}(\mathrm{Mdl}(\mathrm{Idl}(K))). \tag{6}$$

Taking $I = \mathrm{Idl}(K)$ in $\mathrm{Idl}(\mathrm{Mdl}(I)) \subseteq I$ we get

$$\mathrm{Idl}(\mathrm{Mdl}(\mathrm{Idl}(K))) \subseteq \mathrm{Idl}(K)$$

which together with (6) establishes the second claim.

6) Obviously,

$$
\begin{aligned}
\mathrm{Idl}(K_1 + K_2) &= \sum_{g \in \mathrm{Hom}_R(M,A)} (K_1 + K_2)g \\
&= \sum_{g \in \mathrm{Hom}_R(M,A)} K_1 g + \sum_{g \in \mathrm{Hom}_R(M,A)} K_2 g \\
&= \mathrm{Idl}(K_1) + \mathrm{Idl}(K_2).
\end{aligned}
$$

7) Equally simply, $\mathrm{Mdl}(I_1 \cap I_2) = \{ f \in H \mid f \, \mathrm{Hom}_R(M,A) \subseteq I_1 \cap I_2 \} = \mathrm{Mdl}(I_1) \cap \mathrm{Mdl}(I_2)$. $\qquad\square$

Remark. In Lemma 6.2 we can rephrase 3) in terms of annihilators. The containment $IH \subseteq \mathrm{Mdl}(I)$ means that

$$I \subseteq \mathrm{Ann}_S(H/\mathrm{Mdl}(I)) \tag{7}$$

and $\mathrm{Idl}(K)H \subseteq K$ means

$$\mathrm{Idl}(K) \subseteq \mathrm{Ann}_S(H/K). \tag{8}$$

We reformulate (some of) our results.

Theorem 6.3. *Let A and M be right R-modules, $H := \mathrm{Hom}_R(A, M)$, $S := \mathrm{End}(M_R)$, and $T := \mathrm{End}(A_R)$. Let $\mathcal{L}(S)$ denote the lattice of all two-sided ideals of S and $\mathcal{L}(H)$ the lattice of all S-T-submodules of H. Then*

$$\mathrm{Mdl} : \mathcal{L}(S) \to \mathcal{L}(H), \quad \mathrm{Idl} : \mathcal{L}(H) \to \mathcal{L}(S)$$

are inclusion preserving maps with

$$\mathrm{Mdl}(\mathrm{Idl}(\mathrm{Mdl}(I))) = \mathrm{Mdl}(I) \ \text{and} \ \mathrm{Idl}(\mathrm{Mdl}(\mathrm{Idl}(K))) = \mathrm{Idl}(K).$$

Proof. Lemma 6.2.   □

Remark. With the standard notation $T := \mathrm{End}(A_R)$ we also have

$$\mathrm{Mdl} : \mathcal{L}(T) \to \mathcal{L}(H) : \mathrm{Mdl}(I) = \{f \in H \mid \mathrm{Hom}(M, A)f \subseteq I\}$$

where I is an ideal of T, and

$$\mathrm{Idl} : \mathcal{L}(H) \to \mathcal{L}(T) : \mathrm{Idl}(K) = \textstyle\sum_{g \in \mathrm{Hom}_R(M,A)} gK$$

where K is an S-T-submodule of H, with properties as for the previous operators Mdl and Idl. If two-sided ideals of S and two-sided ideals of T are consid-
ered simultaneously we differentiate between $\mathrm{Idl}(K) = \mathrm{Idl}_S(K)$ and $\mathrm{Idl}_T(K) :=$
$\sum_{g \in \mathrm{Hom}_R(M,A)} gK \subseteq T$.

Question 6.4. Natural questions arise:

1) For which A, M, I, and K is it true that $\mathrm{Mdl}(\mathrm{Idl}(K)) = K$ or $\mathrm{Idl}(\mathrm{Mdl}(I)) = I$?

2) When is it true that
   $I\,\mathrm{Hom}_R(A, M) = \mathrm{Mdl}(I)$ or $\mathrm{Idl}(K)\,\mathrm{Hom}_R(A, M) = K$?

3) For which A and M is $\mathrm{Idl}(\mathrm{Rad}(A, M)) = \mathrm{Rad}(S)$?

Example 6.5. Let p be a prime, $A := \mathbb{Z}(p)$ and $M := \mathbb{Z}(p^\infty)$. Then

- $T := \mathrm{End}(A) = \mathbb{Z}/p\mathbb{Z}$, $\mathrm{Rad}(T) = 0$,
- $S := \mathrm{End}(M) = \widehat{\mathbb{Z}}_p$ (the ring of p-adic integers), $\mathrm{Rad}(S) = p\widehat{\mathbb{Z}}_p$,
- $H := \mathrm{Hom}(A, M) = \mathrm{Hom}(A, M[p]) = \mathbb{Z}(p)$,
- $\mathrm{Hom}(M, A) = 0$,
- $\mathrm{Rad}(H) = \mathrm{Rad}(A, M) = \{f \in H \mid f\,\mathrm{Hom}(M, A) \subseteq \mathrm{Rad}(S)\} = H$,
- $\mathrm{Idl}(\mathrm{Rad}(H)) := \sum_{g \in \mathrm{Hom}_R(M,A)} \mathrm{Rad}(A, M)g = 0$.

Hence in this case

$$\mathrm{Idl}(\mathrm{Rad}(H)) \subsetneq \mathrm{Rad}(S).$$

Of particular interest is the equality $I = \mathrm{Idl}(\mathrm{Mdl}(I))$ because it will turn
Mdl Idl into a closure operator as we will see below. We have

$$\mathrm{Idl}(\mathrm{Mdl}(I)) = \textstyle\sum_{g \in \mathrm{Hom}_R(M,A)} \mathrm{Mdl}(I)g \subseteq I.$$

To get equality it is necessary and sufficient to find for each $s \in I$ mappings
$g_i \in \mathrm{Hom}_R(M, A)$ and $f_i \in \mathrm{Mdl}(I)$ such that

$$s = f_1 g_1 + \cdots + f_n g_n.$$

There is an intriguing result pointing in this direction. We ask when can every
endomorphism of M be factored through A? The answer follows.

Lemma 6.6. *Let A, M, and $S = \text{End}(M_R)$ be as before. Then for any $s \in S$ there exist $f \in \text{Hom}_R(A, M)$ and $g \in \text{Hom}_R(M, A)$ such that $s = fg$ if and only if A has a direct summand isomorphic to M.*

Proof. Suppose that $A = M' \oplus N$ and $\sigma : M \to M'$ is an isomorphism. Let $s \in S$. Then let $g : M \to A : g = \sigma s$ and let $f : A \to M$ be such that $f = \sigma^{-1}$ on M' and $f = 0$ on N. Then $fg = s$.

Conversely, assume that $S \ni 1 = fg$ for some $f \in \text{Hom}_R(A, M)$ and some $g \in \text{Hom}_R(M, A)$. Then f is surjective and $A = M' \oplus \text{Ker}(f)$ with $\text{Im}(g) = M' \cong M$. $\square$

Lemma 6.6 suggests specializing to modules A that have a direct summand isomorphic with M.

Lemma 6.7. *Let M be an R-module and $A = M' \oplus N$ with $M' \cong M$. Let I be an ideal of $S := \text{End}(M_R)$. Then, given $s \in I$, there is $\sigma \in \text{Hom}_R(M, A)$ and $f \in \text{Mdl}(I)$ such that $s = f\sigma$.*

Proof. By hypothesis there is an isomorphism $\sigma : M \to M'$ and we let $\pi : A \to M'$ be the projection along N. Let $s \in I$. Let $f : A \to M : f := s\sigma^{-1}\pi$. Then $s = f\sigma$ and we claim that $f \in \text{Mdl}(I)$. This requires that for all $g \in \text{Hom}_R(M, A)$ the composite $fg \in I$. So let $g \in \text{Hom}_R(M, A)$. Then $fg = s\sigma^{-1}\pi g$ with $\sigma^{-1}\pi g \in S$. Since I is an ideal and $s \in I$, we get that $fg \in I$. $\square$

Note that by Lemma 6.7 *the crucial hypothesis* (9) *of the following theorem is satisfied if A has a direct summand isomorphic to M.*

Theorem 6.8. *Let A and M be R-modules, $S := \text{End}(M_R)$, $H := \text{Hom}_R(A, M)$, and assume that*

$$\text{Idl} \, \text{Mdl}(I) = I \text{ for all ideals } I \text{ of } S. \tag{9}$$

Let $\mathcal{L}(S)$ be the set of all ideals of S. Then $\mathcal{L}(S)$ is a complete lattice under set inclusion where the least upper bounds are the sums and the greatest lower bounds are the intersections. Let $\mathcal{L}(H)^{\text{cl}} := \{K \subseteq {}_S H_T \mid K = \text{Mdl} \, \text{Idl}(K)\}$, partially ordered by set inclusion. Then $\text{Mdl}(I) \in \mathcal{L}(H)^{\text{cl}}$ for every $I \in \mathcal{L}(S)$ and $\mathcal{L}(H)^{\text{cl}}$ is a complete lattice where the greatest lower bounds are the intersections. The functions

$$\text{Mdl} : \mathcal{L}(S) \to \mathcal{L}(H)^{\text{cl}} \quad and \quad \text{Idl} : \mathcal{L}(H)^{\text{cl}} \to \mathcal{L}(S)$$

are inverse lattice isomorphisms.

Proof. The lattice properties of $\mathcal{L}(S)$ are well-known and routinely verified.

To show that $\mathcal{L}(H)^{\text{cl}}$ is a complete lattice, we first note that $H \in \mathcal{L}(H)^{\text{cl}}$ because by Lemma 6.2.4), $H \subseteq \text{Mdl} \, \text{Idl}(H) \subseteq H$. We show next that $\mathcal{L}(H)^{\text{cl}}$ is closed under arbitrary intersections which establishes the existence of greatest lower bounds. Let K_i be a family of modules in $\mathcal{L}(H)^{\text{cl}}$. Then $\bigcap K_i \subseteq K_j$ implies that $\text{Idl}(\bigcap K_i) \subseteq \bigcap_j \text{Idl}(K_j)$ and further $\bigcap_j K_j \subseteq \text{Mdl} \, \text{Idl}(\bigcap_j K_j) \subseteq \bigcap_j \text{Mdl} \, \text{Idl}(K_j) = \bigcap_j K_j$. To show the existence of least upper bounds in $\mathcal{L}(H)^{\text{cl}}$, let K_i be a family of modules in $\mathcal{L}(H)^{\mathrm{cl}}$. Let $U := \{K \in \mathcal{L}(H)^{\mathrm{cl}} \mid \forall i : K_i \subseteq K\}$. Then $H \in U$ and $L = \bigcap U \in \mathcal{L}(H)^{\mathrm{cl}}$ because $\mathcal{L}(H)^{\mathrm{cl}}$ is closed under intersections, and L is the least upper bound of the family by the way it is defined.

It follows from Lemma 6.2.6) that $\mathrm{Mdl}(I) \in \mathcal{L}(H)^{\mathrm{cl}}$.

By what has been shown, it is clear the the operators Mdl and Idl are inverse lattice isomorphisms. $\qquad\qquad\square$

7   Correspondences for Modules

We consider now the special case $\mathrm{Hom}_R(R, M) \cong M$. For readers that are only interested in this case, we repeat the main notions and results.

Let us consider an R-module M_R, and $S := \mathrm{End}(M_R)$. Then M is an S-R-bimodule.

Definition 7.1. Let $K \subseteq {}_S M_R$, $I \subseteq {}_S S_S$, and $J \subseteq {}_R R_R$. In particular, I and J are two-sided ideals in S and R respectively. For $m \in M$ and $\psi \in \mathrm{Hom}_R(M, R)$, we have $m\psi : M \to M : (m\psi)(x) = m \cdot \psi(x)$, while for $m \in M$ and $\varphi \in \mathrm{Hom}_R(M, R)$, we have $\varphi m : R \to R : (\varphi m)(x) = \varphi(m) \cdot x$. Hence $m\psi \in S$ and $\varphi m \in \mathrm{End}(R_R) = R$. By specializing the general concepts we arrive at the following definitions.

1) $\mathrm{Idl}_S(K) := \sum_{\psi \in \mathrm{Hom}_R(M,R)} K\psi \subseteq S$, and
   $\mathrm{Idl}_R(K) := \sum_{\varphi \in \mathrm{Hom}_R(M,R)} \varphi K$.

2) $\mathrm{Mdl}(I) := \{m \in M \mid m \, \mathrm{Hom}_S(M, S) \subseteq I\}$, and
   $\mathrm{Mdl}(J) := \{m \in M \mid \mathrm{Hom}_R(M, R) m \subseteq J\}$.

Remark. The map $m\psi : M \to M : (m\psi)(x) = m \cdot \psi(x)$ is an S-homomorphism because for $s \in S$ and $x \in M$,

$$\begin{aligned}((sm)\psi)(x) &= s(m)\psi(x) = s(m\psi(x)) \\ &= s((m\psi)(x)) = (s \circ (m\psi))(x) = (s(m\psi))(x),\end{aligned}$$

so $(sm)\psi = s(m\psi)$. Here we used that $s \cdot m = s(m)$.

Corollary 7.2. Let $\mathcal{L}(S)$ denote the lattice of all two-sided ideals of S and $\mathcal{L}(M)$ the lattice of all S-R-submodules of M. Then

$$\mathrm{Mdl} : \mathcal{L}(S) \to \mathcal{L}(M), \quad \mathrm{Idl} : \mathcal{L}(M) \to \mathcal{L}(S),$$

$$\mathrm{Mdl} : \mathcal{L}(R) \to \mathcal{L}(M), \quad \mathrm{Idl} : \mathcal{L}(M) \to \mathcal{L}(R)$$

are well-defined inclusion preserving maps with the following additional properties where $K \subseteq {}_S M_R$, I is a two-sided ideal of S, and J is a two-sided ideal of R.

1) $K \subseteq \mathrm{Mdl}(\mathrm{Idl}_S(K))$, $\quad K \subseteq \mathrm{Mdl}(\mathrm{Idl}_R(K))$,
   $\mathrm{Idl}_S(\mathrm{Mdl}(I)) \subseteq I$, $\quad \mathrm{Idl}_R(\mathrm{Mdl}(J)) \subseteq J$.

2) $IM \subseteq \mathrm{Mdl}(I)$, and $MJ \subseteq \mathrm{Mdl}(J)$.

3) $\mathrm{Mdl}(\mathrm{Idl}_S(\mathrm{Mdl}(I))) = \mathrm{Mdl}(I)$, $\mathrm{Mdl}(\mathrm{Idl}_R(\mathrm{Mdl}(J))) = \mathrm{Mdl}(J)$.

4) $\mathrm{Idl}_S(\mathrm{Mdl}(\mathrm{Idl}_S(K))) = \mathrm{Idl}_S(K)$, $\mathrm{Idl}_R(\mathrm{Mdl}(\mathrm{Idl}_R(K))) = \mathrm{Idl}_R(K)$.

5) *For S-R-submodules K_1, K_2 of M,*

$$\mathrm{Idl}(K_1 + K_2) = \mathrm{Idl}(K_1) + \mathrm{Idl}(K_2).$$

6) *For ideals I_1, I_2 of S,*

$$\mathrm{Mdl}(I_1 \cap I_2) = \mathrm{Mdl}(I_1) \cap \mathrm{Mdl}(I_2).$$

Proof. All claims are special cases of the general results, in particular Lemma 6.2 but can also be derived analogously to the general results. As an example we prove that $IM \subseteq \mathrm{Mdl}(I)$. Let $s \in I$, $m \in M$, $\varphi \in \mathrm{Hom}_R(M, R)$. Then

$$(sm)\varphi = s(m\varphi) \in I,$$

since $m\varphi \in R$ and I is a right ideal in R, hence $sm \in \mathrm{Mdl}(I)$. □

We specialize Lemma 6.7.

Lemma 7.3. *Let M be an R-module and $R = M' \oplus N$ with $M' \cong M$. Let I be an ideal of $S := \mathrm{End}(M_R)$. Then, given $s \in I$, there is $\sigma \in \mathrm{Hom}_R(M, R)$ and $f \in \mathrm{Mdl}(I)$ such that $s = f\sigma$.*

This shows that the assumption (10) is satisfied if M is isomorphic to a direct summand of R.

Theorem 7.4. *Let M be an R-module, $S := \mathrm{End}(M_R)$, and assume that*

$$\mathrm{Idl}_S \mathrm{Mdl}(I) = I \text{ for all ideals } I \text{ of } S. \tag{10}$$

Let $\mathcal{L}(S)$ be the set of all two-sided ideals of S. Then $\mathcal{L}(S)$ is a complete lattice under set inclusion where the least upper bounds are the sums and the greatest lower bounds are the intersections. Let $\mathcal{L}(M)^{\mathrm{cl}} := \{K \subseteq {}_S M_R \mid K = \mathrm{Mdl}\,\mathrm{Idl}_S(K)\}$, partially ordered by set inclusion. Then $\mathrm{Mdl}(I) \in \mathcal{L}(H)^{\mathrm{cl}}$ for every $I \in \mathcal{L}(S)$ and $\mathcal{L}(M)^{\mathrm{cl}}$ is a complete lattice where the greatest lower bounds are the intersections. The functions

$$\mathrm{Mdl} : \mathcal{L}(S) \to \mathcal{L}(M)^{\mathrm{cl}} \quad and \quad \mathrm{Idl}_S : \mathcal{L}(M)^{\mathrm{cl}} \to \mathcal{L}(S)$$

are inverse lattice isomorphisms.

Proof. This is just a special case of Theorem 6.8. □

Similarly we have the following result.

Theorem 7.5. *Let M be an R-module, $S := \mathrm{End}(M_R)$, and assume that*

$$\mathrm{Idl}_R\,\mathrm{Mdl}(J) = J \text{ for all ideals } J \text{ of } R. \tag{11}$$

Let $\mathcal{L}(R)$ be the set of all two-sided ideals of R. Then $\mathcal{L}(R)$ is a complete lattice under set inclusion where the least upper bounds are the sums and the greatest lower bounds are the intersections. Let $\mathcal{L}(M)^{\mathrm{cl}} := \{K \subseteq {}_S M_R \mid K = \mathrm{Mdl}\,\mathrm{Idl}_R(K)\}$, partially ordered by set inclusion. Then $\mathrm{Mdl}(J) \in \mathcal{L}(H)^{\mathrm{cl}}$ for every $J \in \mathcal{L}(R)$ and $\mathcal{L}(M)^{\mathrm{cl}}$ is a complete lattice where the greatest lower bounds are the intersections. The functions

$$\mathrm{Mdl} : \mathcal{L}(R) \to \mathcal{L}(M)^{\mathrm{cl}} \quad and \quad \mathrm{Idl}_R : \mathcal{L}(M)^{\mathrm{cl}} \to \mathcal{L}(R)$$

are inverse lattice isomorphisms.

Chapter VIII

Regularity in Homomorphism Groups of Abelian Groups

1 Introduction

We are concerned in this chapter with regularity in homomorphism groups of abelian groups. Previous authors [29], [12], [30], [13] dealt with those groups that have regular endomorphism rings. Cognizant of the existence of the largest regular ideal $\mathrm{Reg}(A, A)$ in the endomorphism ring of the group A, their results have been generalized in [24] to computing $\mathrm{Reg}(A, A) = \mathrm{Reg}(\mathrm{End}(A))$. Here we study $\mathrm{Hom}(A, M)$ as an $\mathrm{End}(M)$-$\mathrm{End}(A)$-bimodule in view of regularity. In this chapter "group" will mean "abelian group" and p will always stand for a prime number.

The theory of abelian groups is a highly developed subject so that conclusive answers are possible using the available tools. It is hoped that the results obtained and the ideas employed will be helpful in other contexts.

It is routine to generalize the results of this section to modules over a principal ideal domain.

2 $\mathrm{Hom}(A, M)$ and Regularity

We begin by considering indecomposable groups. The indecomposable torsion groups are completely known. They are

- the cyclic primary groups $\mathbb{Z}(p^n)$ of order p^n, $n \in \mathbb{N}$,

- the divisible Prüfer groups $\mathbb{Z}(p^\infty)$.

In addition *there are very many indecomposable torsion-free groups, among them the rank-one groups, i.e., the additive subgroups of* $\mathbb{Q}$, *in particular* $\mathbb{Q}$ *itself.* There are $2^{\aleph_0}$ non-isomorphic rank-one groups, and there are indecomposable torsion-free groups of any cardinality less than the first strongly inaccessible cardinal ([11, Theorem 89.2]).

Proposition 2.1.

1) $\operatorname{Reg}(\mathbb{Z}(p), \mathbb{Z}(p)) = \operatorname{Hom}(\mathbb{Z}(p), \mathbb{Z}(p)) \cong \mathbb{Z}(p)$.

2) *For $2 \leq n \leq \infty$ and $1 \leq m \leq \infty$,*
   $\operatorname{Reg}(\mathbb{Z}(p^m), \mathbb{Z}(p^n)) = 0$ and $\operatorname{Reg}(\mathbb{Z}(p^n), \mathbb{Z}(p^m)) = 0$.

3) $\operatorname{Reg}(\mathbb{Q}, \mathbb{Q}) = \operatorname{Hom}(\mathbb{Q}, \mathbb{Q}) \cong \mathbb{Q}$.

4) *If A is torsion-free and indecomposable but $A \neq \mathbb{Q}$, then for any group M,*
   $\operatorname{Reg}(A, M) = 0$ and $\operatorname{Reg}(M, A) = 0$.

Proof. 1) is true because $\mathbb{Z}(p)$ is simple and 3) is true because every nonzero map in $\operatorname{Hom}(\mathbb{Q}, \mathbb{Q})$ is an isomorphism.

It is well-known that $\operatorname{End}(\mathbb{Z}(p^n)) \cong \mathbb{Z}/p^n\mathbb{Z}$, and $\operatorname{End}(\mathbb{Z}(p^\infty)) \cong \hat{\mathbb{Z}}_p$, the ring of p-adic integers, and for $n \geq 2$ these endomorphism rings are not division rings. Hence Theorem III.1.1 establishes 2).

4) Suppose that $\operatorname{Reg}(A, M) \neq 0$ or $\operatorname{Reg}(M, A) \neq 0$. Then it follows from Theorem III.1.1 that $\operatorname{End}(A)$ is a division ring. Since (multiplication by) $n \in \mathbb{N}$ is an endomorphism of A, we also have $1/n \in \operatorname{End}(A)$ and this means that A is divisible. The only divisible indecomposable group is $\mathbb{Q}$ (up to isomorphism). We established the contrapositive of the claim. $\square$

In some cases the absence of any nonzero regular maps is easily seen.

Lemma 2.2. *Let A and M be groups such that one of these is torsion and the other is torsion-free. Then $\operatorname{Hom}(A, M)$ contains no regular elements except for the zero map.*

Proof. A direct summand of a torsion group is torsion and a direct summand of a torsion-free group is torsion-free. Hence there is no nonzero summand of A that is isomorphic to a summand of M. By Corollary II.1.5 there are no nonzero regular elements. $\square$

The following lemma says that in the study of torsion groups one can – as is usually done – restrict oneself to primary groups.

Lemma 2.3. *Let A and M be torsion groups and let $A = \bigoplus_{p \in \mathbb{P}} A_p$ and $M = \bigoplus_{p \in \mathbb{P}} M_p$ be the primary decompositions. Then*

$$\operatorname{Hom}(A, M) \cong \prod_{p \in \mathbb{P}} \operatorname{Hom}(A_p, M_p), \text{ and } \operatorname{Reg}(A, M) \cong \prod_{p \in \mathbb{P}} \operatorname{Reg}(A_p, M_p).$$

Proof. It is well-known and easy to see (orders of elements!) that $\text{Hom}(A, M) \cong \prod_{p \in \mathbb{P}} \text{Hom}(A_p, M_p)$. Similarly, $\text{End}(A) \cong \prod_{p \in \mathbb{P}} \text{End}(A_p)$ and $\text{End}(M) \cong \prod_{p \in \mathbb{P}} \text{End}(M_p)$. The actions of $(\alpha_p) \in \text{End}(A)$ and of $(\mu_p) \in \text{End}(M)$ on $(f_p) \in \text{Hom}(A, M)$ is given by

$$(\mu_p)(f_p)(\alpha_p) = (\mu_p f_p \alpha_p),$$

i.e., the action is componentwise, and the claim is now clear. $\qquad\square$

Proposition 2.4. *Let A and M be divisible torsion groups. Then*

$$\text{Reg}(A, M) = 0.$$

Proof. By Lemma 2.3 we assume without loss of generality that A and M are p-groups. Divisible p-groups are direct sums of groups isomorphic to $\mathbb{Z}(p^\infty)$. Suppose that $0 \neq f \in \text{Reg}(A, M)$. Then there are decompositions $A = A_1 \oplus A_2$ and $M = M_1 \oplus M_2$ such that $A_1 \cong M_1 \cong \mathbb{Z}(p^\infty)$ and $0 \neq f(A_1) \subseteq M_1$. By Theorem II.6.14 the map f induces a map $0 \neq f_{11} \in \text{Reg}(A_1, M_1)$ which is a contradiction by Proposition 2.1. $\qquad\square$

It is well-known that every group A contains a unique largest divisible subgroup D_A and since the divisible abelian groups are exactly the injective groups, we have a direct decomposition $A = D_A \oplus R_A$, where R_A contains no nonzero divisible subgroup. Groups that contain no nonzero divisible subgroups are called **reduced**.

Lemma 2.5. *Let A and M be groups such that one of these is divisible and the other is reduced. Then $\text{Hom}(A, M)$ contains no nonzero regular maps.*

Proof. A direct summand of a reduced group is reduced and a direct summand of a divisible group is divisible. Hence there is no nonzero summand of A that is isomorphic to a summand of M. By Corollary II.1.5 the group $\text{Hom}(A, M)$ contains no nonzero regular element. $\qquad\square$

We show next that for the purpose of computing $\text{Reg}(A, M)$ we may assume that both A and M are divisible or that both are reduced.

Lemma 2.6. *Let A and M be any groups and consider decompositions $A = D_A^{\text{tf}} \oplus D_A^{\text{t}} \oplus R_A$ and $M = D_M^{\text{tf}} \oplus D_M^{\text{t}} \oplus R_M$ where $D_A^{\text{t}}, D_M^{\text{t}}$ are divisible torsion groups, $D_A^{\text{tf}}, D_M^{\text{tf}}$ are divisible torsion-free groups, and R_A, R_M are reduced groups. Such decompositions exist. Let $f \in \text{Hom}(A, M)$ and represent f as*

$$f = \begin{bmatrix} f_{11} & f_{21} & f_{31} \\ f_{12} & f_{22} & f_{32} \\ f_{13} & f_{23} & f_{33} \end{bmatrix} \quad \textit{where}$$

$$\begin{array}{lll} f_{11} \in \text{Hom}(D_A^{\text{tf}}, D_M^{\text{tf}}), & f_{21} \in \text{Hom}(D_A^{\text{t}}, D_M^{\text{tf}}), & f_{31} \in \text{Hom}(R_A, D_M^{\text{tf}}), \\ f_{12} \in \text{Hom}(D_A^{\text{tf}}, D_M^{\text{t}}), & f_{22} \in \text{Hom}(D_A^{\text{t}}, D_M^{\text{t}}), & f_{32} \in \text{Hom}(R_A, D_M^{\text{t}}), \\ f_{13} \in \text{Hom}(D_A^{\text{tf}}, R_M), & f_{23} \in \text{Hom}(D_A^{\text{t}}, R_M), & f_{33} \in \text{Hom}(R_A, R_M). \end{array}$$

Then the following statements are true.

1) If $f \in \text{Reg}(A, M)$, then $f_{21} = 0$, $f_{31} = 0$, $f_{12} = 0$, $f_{22} = 0$, $f_{32} = 0$, $f_{13} = 0$, and $f_{23} = 0$.

2) If $f \in \text{Reg}(A, M)$ and $D_M^t \neq 0$, then $f_{11} = 0$.

3) If $f \in \text{Reg}(A, M)$ and $R_A/\text{t}(R_A) \neq 0$, then $f_{11} = 0$.

4) Assume that $D_M^t = 0$ and $R_A/\text{t}(R_A) = 0$. Then

$$f = \begin{bmatrix} f_{11} & 0 & 0 \\ 0 & 0 & 0 \\ 0 & 0 & 0 \end{bmatrix} \in \text{Reg}(A, M).$$

5) If $f \in \text{Reg}(A, M)$, then $\text{Im}(f_{33}) \subseteq \bigoplus \{(R_M)_p \mid (D_m^t)_p = 0\} \subseteq \text{t}(R_M)$.

6) Assume that $\text{Im}(f_{33}) \subseteq \bigoplus \{(R_M)_p \mid (D_m^t)_p = 0\}$, and $f_{33} \in \text{Reg}(R_A, R_M)$. Then

$$f = \begin{bmatrix} 0 & 0 & 0 \\ 0 & 0 & 0 \\ 0 & 0 & f_{33} \end{bmatrix} \in \text{Reg}(A, M).$$

7) Assume that $D_M^t = 0$, $R_A/\text{t}(R_A) = 0$, $\text{Im}(f_{33}) \subseteq \bigoplus \{(R_M)_p \mid (D_m^t)_p = 0\}$, and $f_{33} \in \text{Reg}(R_A, R_M)$. Then

$$f = \begin{bmatrix} f_{11} & 0 & 0 \\ 0 & 0 & 0 \\ 0 & 0 & f_{33} \end{bmatrix} \in \text{Reg}(A, M).$$

Proof. 1) Assume that $f \in \text{Reg}(A, M)$. Then $f_{22} = 0$ by Theorem II.6.14 and Proposition 2.4, $f_{21} = 0$ because there are no nonzero maps on a torsion group to a torsion-free group, $f_{13} = 0$ and $f_{23} = 0$ because there are no nonzero maps of a divisible group to a reduced group; the other statements are true by Theorem II.6.14, Lemma 2.2 and Lemma 2.5.

2) By Theorem II.6.14 we have $\mu f_{11} \in \text{Reg}(D_A^{tf}, D_M^t) = 0$, for every map $\mu \in \text{Hom}(D_M^{tf}, D_M^t)$. Suppose that $f_{11} \neq 0$. Then $0 \neq \text{Im}(f_{11}) \subseteq D_M^{tf}$ and there is $\mu \in \text{Hom}(D_M^{tf}, D_M^t)$ such that $\mu f_{11} \neq 0$, a contradiction. We conclude that $f_{11} = 0$.

3) By Theorem II.6.14 we have $f_{11}\alpha \in \text{Reg}(R_A, D_M^{tf}) = 0$, for every map $\alpha \in \text{Hom}(R_A, D_M^{tf})$. Suppose that $f_{11} \neq 0$. Then $0 \neq \text{Im}(f_{11}) \subseteq D_M^{tf}$ and, as $R_A/\text{t}(R_A) \neq 0$, there is $\alpha \in \text{Hom}(R_A, D_M^{tf})$ such that $f_{11}\alpha \neq 0$, a contradiction. We conclude that $f_{11} = 0$.

4) By Theorem II.6.14 we have to show that all possible maps of the form $\mu f_{11}\alpha$ where μ has domain D_M^{tf} and α has codomain D_A^{tf} lie in the appropriate

Reg. The possibilities are depicted in the following diagram.

$$
\begin{array}{ccccccc}
D_A^{\text{tf}} & & & & & & D_M^{\text{tf}} \\
& \searrow & & & & \nearrow & \\
D_A^{\text{t}} & \longrightarrow & D_A^{\text{tf}} & \xrightarrow{f_{11}} & D_M^{\text{tf}} & \longrightarrow & D_M^{\text{t}} \\
& \nearrow & & & & \searrow & \\
R_A & & & & & & R_M
\end{array}
$$

We have Hom$(D_A^{\text{t}}, D_A^{\text{tf}}) = 0$, and Hom$(D_M^{\text{tf}}, R_M) = 0$ which settles affirmatively five of the nine possible combinations $\mu f_{11} \alpha$. But also Hom$(R_A, D_M^{\text{tf}}) = 0$ and Hom$(D_M^{\text{tf}}, D_M^{\text{t}}) = 0$ by our assumptions. This settles three more cases affirmatively, and leaves the single composite $\mu f_{11} \alpha : D_A^{\text{tf}} \to D_A^{\text{tf}} \xrightarrow{f_{11}} D_M^{\text{tf}} \to D_M^{\text{tf}}$ which is in Reg$(D_A^{\text{tf}}, D_M^{\text{tf}})$ because Reg$(D_A^{\text{tf}}, D_M^{\text{tf}}) = $ Hom$(D_A^{\text{tf}}, D_M^{\text{tf}})$.

5) Suppose that $f \in$ Reg(A, M) and Im$(f_{33}) \not\subseteq \bigoplus \{(R_M)_p \mid (D_m^{\text{t}})_p = 0\}$. Then there is $\mu : R_M \to D_M^{\text{t}}$ such that $0 \neq \mu f_{33} \in$ Reg$(R_A, D_M^{\text{t}}) = 0$, a contradiction. Hence Im$(f_{33}) \subseteq \bigoplus \{(R_M)_p \mid (D_m^{\text{t}})_p = 0\}$.

6) By Theorem II.6.14 we have to show that all possible maps of the form $\mu f_{33} \alpha$ where μ has domain R_M and α has codomain R_A lie in the appropriate Reg. The possibilities are depicted in the following diagram.

$$
\begin{array}{ccccccc}
D_A^{\text{tf}} & & & & & & D_M^{\text{tf}} \\
& \searrow & & & & \nearrow & \\
D_A^{\text{t}} & \longrightarrow & R_A & \xrightarrow{f_{33}} & R_M & \longrightarrow & D_M^{\text{t}} \\
& \nearrow & & & & \searrow & \\
R_A & & & & & & R_M
\end{array}
$$

We have Hom$(D_A^{\text{tf}}, R_A) = 0$, and Hom$(D_M^{\text{t}}, R_A) = 0$ which settles affirmatively six of the nine possible combinations $\mu f_{33} \alpha$. But also the composite $\mu f_{11} \alpha : R_A \to R_A \xrightarrow{f_{33}} R_M \to D_M^{\text{tf}}$ is equal to 0 because Im$(f_{33}) \subseteq \mathbf{t}(R_M)$ by assumption. Further the composite $\mu f_{11} \alpha : R_A \to R_A \xrightarrow{f_{33}} R_M \to D_M^{\text{t}}$ is equal to 0 because of the assumption that Im$(f_{33}) \subseteq \bigoplus \{(R_M)_p \mid (D_m^{\text{t}})_p = 0\}$. This leaves the case $\mu f_{11} \alpha : R_A \to R_A \xrightarrow{f_{33}} R_M \to R_M$. In this case $\mu f_{11} \alpha \in$ Reg(R_A, R_M) because $f_{33} \in$ Reg(R_A, R_M) by assumption and $\mu \in$ End(R_M) and $\alpha \in$ End(R_A).

7) is the combination of 4) and 6). $\qquad\square$

We wish to extend these results by computing the maximum regular bisubmodule of Hom(A, M) in general. A first step consists in checking when pf is regular for a regular homomorphism $f \in$ Hom$(A.M)$.

Lemma 2.7. Let A and M be abelian groups and suppose that $f \in$ Hom(A, M) is regular. Then pf is also regular if and only if

$$
\text{Im}(f) = p\,\text{Im}(f) \oplus \text{Im}(f)[p]. \tag{1}
$$

Proof. As f is regular we have decompositions

$$A = \mathrm{Ker}(f) \oplus A_1, \quad M = \mathrm{Im}(f) \oplus M_1, \quad A_1 \cong \mathrm{Im}(f).$$

Regarding pf, it is easily seen that

$$\mathrm{Im}(pf) = p\,\mathrm{Im}(f), \quad \text{and} \quad \mathrm{Ker}(pf) = \mathrm{Ker}(f) \oplus A_1[p].$$

Let D denote the maximal divisible subgroup of $\mathrm{Im}(f)$.

Assume first that pf is also regular. Then $\mathrm{Im}(pf) = p\,\mathrm{Im}(f) \subseteq^{\oplus} M$, hence $\mathrm{Im}(f) = p\,\mathrm{Im}(f) \oplus C$ for some C, and $pC \cong p\,(\mathrm{Im}(f)/p\,\mathrm{Im}(f)) = 0$, so $C \subseteq \mathrm{Im}(f)[p]$. Therefore $p\,\mathrm{Im}(f) = p(p\,\mathrm{Im}(f))$. Also $D = pD \subseteq p\,\mathrm{Im}(f)$, and hence $p\,\mathrm{Im}(f) = D \oplus B$ for some subgroup B. Here $pB = B$ as B is a summand of a p-divisible group, but B is reduced by the choice of D. This means that the p-primary component B_p of B must be 0 because p-divisible p-groups are divisible. So

$$(\mathrm{Im}(f))[p] = D[p] \oplus C. \tag{2}$$

At this point we have $A_1 \cong \mathrm{Im}(f) = D \oplus B \oplus C$. Therefore $A_1 = D' \oplus B' \oplus C'$ with $D' \cong D$, $B' \cong B$, and $C' \cong C$. As $\mathrm{Ker}(pf) = \mathrm{Ker}(f) \oplus A_1[p]$ is a summand of A, so is $A_1[p] = D'[p] \oplus B'[p] \oplus C'[p]$. This makes $D'[p]$ a summand of A and it follows from the structure theory of divisible abelian groups that $D'[p] = 0$. Thus $D[p] = 0$ and by (2), $C = (\mathrm{Im}(f))[p]$. Altogether we have $\mathrm{Im}(f) = D \oplus B \oplus C = p\,\mathrm{Im}(f) \oplus \mathrm{Im}(f)[p]$ as claimed.

Assume now that $\mathrm{Im}(f) = p\,\mathrm{Im}(f) \oplus \mathrm{Im}(f)[p]$. We must show that pf is regular and this is the case if $\mathrm{Im}(pf) \subseteq^{\oplus} M$ and $\mathrm{Ker}(pf) \subseteq^{\oplus} A$. As $\mathrm{Im}(pf) = p\,\mathrm{Im}(f)$, it follows from the assumptions and the fact that a summand of a summand is a summand that $\mathrm{Im}(pf) \subseteq^{\oplus} M$. From $A_1 \cong \mathrm{Im}(f)$ it follows from the assumptions that $A_1 = A_1[p] \oplus pA_1$ and then from $A = \mathrm{Ker}(f) \oplus A_1$ that $A = \mathrm{Ker}(f) \oplus A_1[p] \oplus pA_1$. Since $\mathrm{Ker}(pf) = \mathrm{Ker}(f) \oplus A_1[p]$ as observed above, we have $A = \mathrm{Ker}(pf) \oplus pA_1$ as needed. $\square$

Corollary 2.8. *Suppose that $f \in \mathrm{Hom}(A, M)$ and pf are both regular. Then $p^k f$ is regular for any integer $k \geq 0$.*

Proof. By Lemma 2.7 we have $A = \mathrm{Ker}(f) \oplus A_1$, $M = \mathrm{Im}(f) \oplus M_1$ and $\mathrm{Im}(f) = p\,\mathrm{Im}(f) \oplus \mathrm{Im}(f)[p]$. We will show that $\mathrm{Im}(p^k f) \subseteq^{\oplus} M$ and $\mathrm{Ker}(p^k f) \subseteq^{\oplus} A$ which says that $p^k f$ is regular. The group $p\,\mathrm{Im}(f)$ is p-divisible and hence for any $k \geq 1$, we have $\mathrm{Im}(p^k f) = p^{k-1}(p\,\mathrm{Im}(f)) = p\,\mathrm{Im}(f)$ and this is a direct summand of M. It is easily seen that $\mathrm{Ker}(p^k f) = \mathrm{Ker}(f) \oplus A_1[p^k]$. Since $(p\,\mathrm{Im}(f))_p = 0$, we have that $\mathrm{Im}(f)[p^k] = \mathrm{Im}(f)[p]$ which is a direct summand of $\mathrm{Im}(f)$. As $A_1 \cong \mathrm{Im}(f)$, also $A_1[p^k] = A_1[p]$ is a direct summand of A_1 and hence $\mathrm{Ker}(p^k f) = \mathrm{Ker}(f) \oplus A_1[p^k]$ is a direct summand of A which concludes the proof. $\square$

In the same vein we characterize regular cyclic subgroups of $\mathrm{Hom}(A, M)$.

Proposition 2.9. *Let A be a group and let $f \in \text{Hom}(A, M)$. Then $\mathbb{Z}f := \{nf \mid n \in \mathbb{Z}\}$ is a regular subgroup of $\text{Hom}(A, M)$ if and only if*

$$\text{Ker}(f) \subseteq^{\oplus} A, \quad \text{and} \quad \text{Im}(f) \subseteq^{\oplus} M \tag{3}$$

and for all $n \in \mathbb{Z}$,

$$\text{Im}(f) = n\,\text{Im}(f) \oplus \text{Im}(f)[n]. \tag{4}$$

Proof. Suppose that $\mathbb{Z}f$ is regular. Then f is regular and there exist decompositions

$$A = \text{Ker}(f) \oplus A_1, \quad M = \text{Im}(f) \oplus M_1. \tag{5}$$

We will use several times the induced isomorphism

$$\tilde{f} := f \restriction_{A_1} : A_1 \to \text{Im}(f).$$

Let $n \in \mathbb{Z}$ be given. Then nf is also regular. We first observe that

$$\text{Im}(nf) = n\,\text{Im}(f) \subseteq \text{Im}(f), \quad \text{and} \quad \text{Ker}(nf) = \text{Ker}(f) \oplus A_1[n]. \tag{6}$$

The first of these is obvious. To verify the second, let $x \in A$. Using (5) write $x = y + z$ with $y \in \text{Ker}(f)$ and $z \in A_1$. Then

$$x \in \text{Ker}(nf) \quad \Leftrightarrow \quad (nf)(y + z) = nf(z) = 0 \Leftrightarrow f(z) \in \text{Im}(f)[n]$$
$$\Leftrightarrow \quad z \in A_1[n],$$

where we applied the isomorphism $(\tilde{f})^{-1}$. We verify next that

$$\text{Im}(f) = n\,\text{Im}(f) + \text{Im}(f)[n]. \tag{7}$$

In fact, as $\text{Im}(nf) = n\,\text{Im}(f) \subseteq^{\oplus} M$, there is a direct decomposition $\text{Im}(f) = n\,\text{Im}(f) \oplus M_2$, and $nM_2 \subseteq n\,\text{Im}(f) \cap M_2 = 0$, showing that $M_2 \subseteq \text{Im}(f)[n]$ and so $\text{Im}(f) = n\,\text{Im}(f) \oplus M_2 \subseteq n\,\text{Im}(f) + \text{Im}(f)[n] \subseteq \text{Im}(f)$ and (7) is established.

We will use next that

$$A_1[n] \subseteq^{\oplus} \text{Ker}(nf) \subseteq^{\oplus} A, \quad \text{hence} \quad A_1[n] \subseteq^{\oplus} A_1.$$

Let $A_1 = A_1[n] \oplus A_2$. Then

$$\text{Im}(f) = \tilde{f}(A_1) = \text{Im}(f)[n] \oplus \tilde{f}(A_2). \tag{8}$$

We will now show that $\tilde{f}(A_2) = n\,\text{Im}(f)$ which will establish (4). Let $x \in f(A_2)$. By (7) we can write $x = ny + z$ where $y \in \text{Im}(f)$ and $z \in \text{Im}(f)[n]$. It follows that $nx = n^2 y \in f(A_2) \cap n^2\,\text{Im}(f) = n^2 f(A_2)$ because $f(A_2)$ is pure (even a summand) in $\text{Im}(f)$. Hence there is $x' \in f(A_2)$ such that $nx = n^2 x'$, so $n(x - nx') = 0$. Therefore $x - nx' \in \tilde{f}(A_2) \cap \text{Im}(f)[n] = 0$ by (8), $x = nx' \in nf(A_2)$, and we have shown that $\tilde{f}(A_2) = nf(A_2)$ and further $A_2 = nA_2$. From $A_1 = A_1[n] \oplus A_2$ we now find that $A_2 = nA_2 = nA_1$ and so $A_1 = A_1[n] \oplus nA_1$. Applying the isomorphism $(\tilde{f})^{-1}$ we get $\mathrm{Im}(f) = \mathrm{Im}(f)[n] \oplus n\,\mathrm{Im}(f)$.

For the converse assume that (4) and (3) hold. Then f is regular by (3) and Corollary II.1.3. Let $n \in \mathbb{Z}$. By (4) $\mathrm{Im}(nf) = n\,\mathrm{Im}(f) \subseteq^{\oplus} \mathrm{Im}(f) \subseteq^{\oplus} M$, so $\mathrm{Im}(nf) \subseteq^{\oplus} M$. Also $A = \mathrm{Ker}(f) \oplus A_1$ for some A_1. Furthermore, $A_1 \cong \mathrm{Im}(f) = \mathrm{Im}(f)[n] \oplus n\,\mathrm{Im}(f)$ which says that $A_1[n] \subseteq^{\oplus} A_1$ and we see that $\mathrm{Ker}(nf) = \mathrm{Ker}(f) \oplus A_1[n]$ is a summand of A. By Corollary II.1.3 nf is regular. □

Corollary 2.10.

1) *Suppose that M has no elementary direct summands and either A or M is reduced. Then $\mathrm{Reg}(A, M) = 0$.*

2) *Suppose that A has no elementary direct summands and either A or M is reduced. Then $\mathrm{Reg}(A, M) = 0$.*

Proof. 1) Suppose that $f \in \mathrm{Reg}(A, M)$ and let $p \in \mathbb{P}$. Then pf is regular. By Lemma 2.7 we have $\mathrm{Im}(f) = p\,\mathrm{Im}(f) \oplus \mathrm{Im}(f)[p]$. Also $\mathrm{Im}(f) \subseteq^{\oplus} M$, so $\mathrm{Im}(f)[p] \subseteq^{\oplus} M$, and by hypothesis $\mathrm{Im}(f)[p] = 0$. So $\mathrm{Im}(f) = p\,\mathrm{Im}(f)$ which makes the $\mathrm{Im}(f)$ p-divisible for every prime p which means that $\mathrm{Im}(f)$ is divisible. If M is reduced, it follows that $\mathrm{Im}(f) = 0$ and so $f = 0$. Also $A \cong \mathrm{Ker}(f) \oplus \mathrm{Im}(f)$ (Theorem II.1.2) and if A is reduced then again $\mathrm{Im}(f) = 0$, $f = 0$.

2) We have $A \cong \mathrm{Ker}(f) \oplus \mathrm{Im}(f)$ and for every prime p, $\mathrm{Im}(f) = p\,\mathrm{Im}(f) \oplus \mathrm{Im}(f)[p]$. Since A has no elementary direct summands, $\mathrm{Im}(f)$ is divisible as before. Therefore $\mathrm{Im}(f) = 0$, i.e., $f = 0$, if either A or M is reduced. □

The following useful lemma is just a specialization of Theorem II.6.14 and Corollary II.1.5.

Lemma 2.11. *Suppose that $A = A_1 \oplus A_2$ and $M = M_1 \oplus M_2$ and no nonzero summand of A_1 is isomorphic to a summand of M_2. Suppose that*

$$f = \begin{bmatrix} f_{11} & f_{21} \\ f_{12} & f_{22} \end{bmatrix} \in \mathrm{Reg}(A, M).$$

Then $f_{21} = 0$, $f_{12} = 0$, and for every $\mu_{ij} \in \mathrm{Hom}(M_i, M_j)$, and every $\alpha_{ij} \in \mathrm{Hom}(A_i, A_j)$, where $i, j \in \{1, 2\}$,

$$\mu_{21} f_{22} = 0, \quad \mu_{12} f_{22} = 0, \quad f_{22} \alpha_{12} = 0, \quad f_{11} \alpha_{21} = 0, \quad \text{and}$$

$$f_{11} \in \mathrm{Reg}(A_1, M_1), \quad \text{and} \quad f_{22} \in \mathrm{Reg}(A_2, M_2).$$

We begin with a lemma that will considerably simplify and shorten future proofs.

Lemma 2.12. *Let A and M be abelian groups and let $f \in \mathrm{Reg}(A, M)$. Then $A = \mathrm{Ker}(f) \oplus A_1$ and $M = \mathrm{Im}(f) \oplus M_1$ for subgroups A_1 and M_1. Furthermore,*

1) *for any $\mu \in \mathrm{Hom}(\mathrm{Im}(f), M_1)$, necessarily $\mathrm{Im}(\mu) \subseteq^{\oplus} M_1$ and $\mathrm{Ker}(\mu) \subseteq^{\oplus} \mathrm{Im}(f)$, and*

2) *for any* $\alpha \in \text{Hom}(\text{Ker}(f), A_1)$, *necessarily* $\text{Ker}(\alpha) \subseteq^{\oplus} \text{Ker}(f)$ *and* $\text{Im}(\alpha) \subseteq^{\oplus}$ A_1.

Proof. We will again use the isomorphism

$$\tilde{f}: A_1 \ni x \mapsto f(x) \in \text{Im}(f).$$

Since $f \in \text{Reg}(A, M)$ we also have that, for every $t \in \text{End}(A)$ and every $s \in \text{End}(M)$, the composites ft and sf are regular which means that the images and kernels of such maps are direct summands.

1) Let $\mu \in \text{Hom}(\text{Im}(f), M_1)$. The map μ extends to an endomorphism s of M by setting $s \restriction_{\text{Im}(f)} = \mu$ and $s(M_1) = 0$. Then $\text{Im}(sf) = s(\text{Im}(f)) = \text{Im}(\mu)$ and, sf being regular, this must be a direct summand of M and hence of M_1. Next, we claim that

$$\text{Ker}(sf) = \text{Ker}(f) \oplus \tilde{f}^{-1}(\text{Ker}(\mu)).$$

For the proof let $x \in A$ and write $x = y + z$ where $y \in \text{Ker}(f)$ and $z \in A_1$. Then

$$
\begin{aligned}
x \in \text{Ker}(sf) \quad &\Leftrightarrow \quad 0 = sf(x) = sf(z) \\
&\Leftrightarrow \quad f(z) \in \text{Ker}(\mu) \\
&\Leftrightarrow \quad z \in \tilde{f}^{-1}(\text{Ker}(\mu)).
\end{aligned}
$$

The kernel $\text{Ker}(sf)$ must be a summand of A, and hence $\tilde{f}^{-1}(\text{Ker}(\mu))$ is a direct summand of A_1, therefore $\text{Ker}(\mu)$ a summand of $\text{Im}(f)$. This establishes 1).

2) Let $\alpha \in \text{Hom}(\text{Ker}(f), A_1)$. The map α extends to a endomorphism t of A by setting $t \restriction_{\text{Ker}(f)} = \alpha$ and $t(A_1) = 0$. Then ft must be regular. Therefore $\text{Im}(ft) = f(\text{Im}(t)) \subseteq^{\oplus} \text{Im}(f)$ and, applying $\tilde{f}^{-1}$, $\text{Im}(\alpha) = \text{Im}(t) \subseteq^{\oplus} A_1$. It is easy to see that $\text{Ker}(ft) = \text{Ker}(\alpha) \oplus A_1$ and $\text{Ker}(ft)$ must be a summand of A, hence also $\text{Ker}(\alpha)$ is a summand of A, and this makes $\text{Ker}(\alpha)$ a summand of $\text{Ker}(f)$ as claimed. $\qquad\square$

Remark. In connection with Lemma 2.11 we repeatedly use the following fact. Suppose that E is an elementary p-group, $0 \neq \xi \in \text{End}(E)$ and K a group with $K/pK \neq 0$. Then there exists $\gamma \in \text{Hom}(K, E)$ such that $\xi\gamma \neq 0$. This is true because γ is necessarily a composite of the natural epimorphism $K \twoheadrightarrow K/pK$ with a map $K/pK \to E$, and K/pK and E are just vector spaces over the prime field $\mathbb{Z}/p\mathbb{Z}$ so that maps $K/pK \to E$ are available whose image is not contained in $\text{Ker}(\xi)$.

We have established the machinery to settle the torsion-free and torsion cases.

Theorem 2.13. *Let A and M be torsion-free abelian groups. Then $\text{Reg}(A, M) = 0$ unless A and M are both divisible. If both A and M are divisible, then $\text{Reg}(A, M) = \text{Hom}(A, M)$.*

Proof. Let $0 \neq f \in \operatorname{Reg}(A, M)$. Then $\mathbb{Z}f$ is regular and by Lemma 2.9 we have a decomposition $\operatorname{Im}(f) = n\operatorname{Im}(f) \oplus \operatorname{Im}(f)[n]$. As M is torsion-free, $\operatorname{Im}(f) = n\operatorname{Im}(f)$ which says that $\operatorname{Im}(f)$ is divisible and $M = \operatorname{Im}(f) \oplus D \oplus M_1$ where D is divisible and M_1 is reduced. If $M_1 \neq 0$, then there is $\mu : M_1 \to \operatorname{Im}(f)$ and by Lemma 2.12 it follows that $\operatorname{Im}(\mu) \neq 0$ is a summand of $\operatorname{Im}(f)$, so divisible and so $M_1/\operatorname{Ker}(\mu) \cong \operatorname{Im}(\mu)$ is also divisible but also isomorphic to a nonzero direct summand of M_1. This contradiction shows that M is divisible.

We also have that $A = \operatorname{Ker}(f) \oplus A_1$ where $A_1 \cong \operatorname{Im}(f)$ and we have seen that $\operatorname{Im}(f)$ is divisible. There is a decomposition $\operatorname{Ker}(f) = K_1 \oplus K_2$ such that K_1 is divisible and K_2 is reduced. If $K_2 \neq 0$, then there is $0 \neq \alpha \in \operatorname{Hom}(K_2, \operatorname{Im}(f))$ and by Lemma 2.12 it follows that $0 \neq \operatorname{Im}(\alpha) \subseteq^{\oplus} \operatorname{Im}(f)$ and hence $\operatorname{Im}(\alpha)$ is divisible. But $\operatorname{Im}(\alpha) \cong K_2/\operatorname{Ker}(\alpha)$ is also isomorphic to a direct summand of the reduced group K_2 which is a contradiction. Hence $K_2 = 0$ and A is divisible.

Suppose now that both A and M are divisible. Then both A and M are $\mathbb{Q}$-vector spaces, so $\operatorname{Hom}(A, M) = \operatorname{Hom}_{\mathbb{Q}}(A, M)$ and $\operatorname{Reg}(A, M) = \operatorname{Hom}(A, M)$. $\square$

By Lemma 2.3 there is no loss of generality in assuming that torsion groups are primary.

Theorem 2.14. *Let A and M be p-primary groups.*

1) *Suppose that M is not reduced. Then $\operatorname{Reg}(A, M) = 0$.*

2) *Suppose that M is reduced. There are decompositions $A = A_1 \oplus A_2$ and $M = M_1 \oplus M_2$ such that A_1 and M_1 are elementary, and A_2 and M_2 have no direct summands of order p. Then*

   i) $\operatorname{Reg}(A, M) = 0$ *if $M_2 \neq 0$,*

   ii) $\operatorname{Reg}(A, M) = 0$ *if A_2 is not divisible,*

   iii) $\operatorname{Reg}(A, M) = \operatorname{Hom}(A, M)$ *if $M_2 = 0$ and A_2 is divisible, i.e., if A is the direct sum of an elementary group and a divisible group and M is elementary.*

Proof. 1) Write $A = D_A \oplus R_A$ and $M = D_M \oplus R_M$ such that D_A, D_M are divisible and R_A, R_M are reduced. Let $f \in \operatorname{Hom}(A, M)$. Then

$$f = \begin{bmatrix} f_{DD} & f_{RD} \\ f_{DR} & f_{RR} \end{bmatrix}, \qquad \begin{array}{ll} f_{DD} \in \operatorname{Hom}(D_A, D_M), & f_{RD} \in \operatorname{Hom}(R_A, D_M), \\ f_{DR} \in \operatorname{Hom}(D_A, R_M), & f_{RR} \in \operatorname{Hom}(R_A, R_M). \end{array}$$

Suppose that $f \in \operatorname{Reg}(A, M)$. Then $f_{RD} \in \operatorname{Reg}(R_A, D_M) = 0$ by Lemma 2.5, $f_{DR} = 0$ because epimorphic images of divisible groups are divisible, and $f_{DD} \in \operatorname{Reg}(D_A, D_M) = 0$ (Proposition 2.4). Suppose that $0 \neq f_{RR} \in \operatorname{Reg}(R_A, R_D)$. Then $0 \neq \operatorname{Im}(f_{RR}) \subseteq^{\oplus} R_M$ and $D_M \neq 0$ by hypothesis. Hence there is $0 \neq \mu \in \operatorname{Hom}(\operatorname{Im}(f_{RR}), D_M)$. Then $0 \neq \mu f_{RR} \in \operatorname{Reg}(R_A, D_M) = 0$, a contradiction. Hence $f = f_{RR} = 0$.

2) Let $f \in \text{Reg}(A, M)$. Then

$$\text{Im}(f) = p\,\text{Im}(f) \oplus \text{Im}(f)[p].$$

Hence $p\,\text{Im}(f) = p(p\,\text{Im}(f))$ which means that $p\,\text{Im}(f)$ is divisible and thus $p\,\text{Im}(f) = 0$ because M is reduced. So $\text{Im}(f) = \text{Im}(f)[p]$ is an elementary p-group. Suppose there exists $0 \neq \mu \in \text{Hom}(\text{Im}(f), M_2)$. Then Lemma 2.12 implies that $\text{Im}(\mu) \subseteq^\oplus M_2$. But $\text{Im}(\mu)$ is elementary and M_2 has no nonzero elementary summands. This is a contradiction and shows that $f = 0$ if $M_2 \neq 0$. This proves i).

Now suppose that $M = M_1$ is elementary. Then $pA \subseteq \text{Ker}(f)$ and $A = \text{Ker}(f) \oplus A_3$ where $A_3 \cong \text{Im}(f)$ is elementary. There is a decomposition $\text{Ker}(f) = K_1 \oplus K_2$ such that K_1 is elementary and K_2 has no nonzero elementary direct summands. If $K_2/pK_2 \neq 0$, we have a nonzero mapping

$$\alpha : \text{Ker}(f) \overset{\text{proj}}{\twoheadrightarrow} K_2 \overset{\text{nat}}{\twoheadrightarrow} K_2/pK_2 \to A_3.$$

Lemma 2.12 says that $\text{Ker}(\alpha) \subseteq^\oplus \text{Ker}(f)$ and $pA \oplus K_1 \subseteq \text{Ker}(\alpha)$ which makes the complementary direct summand of $\text{Ker}(\alpha)$ in $\text{Ker}(f)$ a nonzero elementary direct summand of K_2. But K_2 has no such direct summands and this contradiction says that $K_2 = pK_2$, i.e., K_2 is divisible. We conclude that $A = K_1 \oplus K_2 \oplus A_3$ is the direct sum of the divisible group K_2 and the elementary group $K_1 \oplus A_3$. This proves ii).

iii) Suppose that A_2 is divisible. Then $\text{Hom}(A, M) = \text{Hom}(A_1, M)$ and every element of $\text{Hom}(A_1, M)$ is regular because A_1 and M are both elementary. $\qquad\square$

We will next tackle the case of a splitting mixed group, reduced or not.

Proposition 2.15. *Let A and M be splitting mixed groups and consider decompositions $A = A_1 \oplus A_2 \oplus A_3$ and $M = M_1 \oplus M_2 \oplus M_3$ where A_1, M_1 are elementary abelian groups, A_2, M_2 are torsion groups without elementary direct summands, and A_3, M_3 are torsion-free groups. Such decompositions exist. Let $f \in \text{Hom}(A, M)$ and represent f as*

$$f = \begin{bmatrix} f_{11} & f_{21} & f_{31} \\ f_{12} & f_{22} & f_{32} \\ f_{13} & f_{23} & f_{33} \end{bmatrix} \quad where$$

$$\begin{array}{lll} f_{11} \in \text{Hom}(A_1, M_1), & f_{21} \in \text{Hom}(A_2, M_1), & f_{31} \in \text{Hom}(A_3, M_1), \\ f_{12} \in \text{Hom}(A_1, M_2), & f_{22} \in \text{Hom}(A_2, M_2), & f_{32} \in \text{Hom}(A_3, M_2), \\ f_{13} \in \text{Hom}(A_1, M_3), & f_{23} \in \text{Hom}(A_2, M_3), & f_{33} \in \text{Hom}(A_3, M_3). \end{array}$$

Then the following statements are true.

1) *If $f \in \text{Reg}(A, M)$, then $f_{21} = 0$, $f_{31} = 0$, $f_{12} = 0$, $f_{22} = 0$, $f_{32} = 0$, $f_{13} = 0$, and $f_{23} = 0$.*

2) If $f \in \mathrm{Reg}(A, M)$ and $f_{11} \neq 0$, then for every p for which $(A_1)_p \neq 0$ necessarily $A_2 = pA_2$, $A_3 = pA_3$ and $M_2[p] = 0$.

3) Assume that $A_2 = pA_2$, $A_3 = pA_3$ and $M_2[p] = 0$ whenever $(A_1)_p \neq 0$, then

$$f = \begin{bmatrix} f_{11} & 0 & 0 \\ 0 & 0 & 0 \\ 0 & 0 & 0 \end{bmatrix} \in \mathrm{Reg}(A, M).$$

4) If $f \in \mathrm{Reg}(A, M)$ and $f_{33} \neq 0$, then A_3 and M_3 are divisible.

5) Assume that A_3 and M_3 are divisible. Then

$$f = \begin{bmatrix} 0 & 0 & 0 \\ 0 & 0 & 0 \\ 0 & 0 & f_{33} \end{bmatrix} \in \mathrm{Reg}(A, M).$$

6) Assume that $A_2 = pA_2$, $A_3 = pA_3$ and $M_2[p] = 0$ whenever $(A_1)_p \neq 0$, and that A_3 and M_3 are divisible. Then

$$f = \begin{bmatrix} f_{11} & 0 & 0 \\ 0 & 0 & 0 \\ 0 & 0 & f_{33} \end{bmatrix} \in \mathrm{Reg}(A, M).$$

Proof. 1) Assume that $f \in \mathrm{Reg}(A, M)$ and use Theorem II.6.14. Then $f_{21} = 0$, $f_{31} = 0$, $f_{12} = 0$, $f_{22} = 0$, and $f_{32} = 0$ by Corollary 1.5, $f_{13} = 0$ and $f_{23} = 0$ because there are no nonzero maps on a torsion group to a torsion-free group.

2) By Theorem II.6.14 we have $\mu f_{11} \in \mathrm{Reg}(A_1, M_2) = 0$, for every map $\mu \in \mathrm{Hom}(M_1, M_2)$. Suppose that $f_{11} \neq 0$. Then $0 \neq \mathrm{Im}(f_{11}) \subseteq M_1$ and if $M_2[p] \neq 0$, then there is $\mu \in \mathrm{Hom}(M_1, M_2)$ such that $\mu f_{11} \neq 0$, a contradiction. We conclude that $M_2[p] = 0$.

Again by Theorem II.6.14 we have $f_{11}\alpha \in \mathrm{Reg}(A_2, M_1) = 0$, for every map $\alpha \in \mathrm{Hom}(A_2, A_1)$. Suppose that $f_{11} \neq 0$. Then $0 \neq \mathrm{Im}(f_{11}) \subseteq M_1$ and, if $A_2/pA_2 \neq 0$, there is $\alpha \in \mathrm{Hom}(A_2, A_1)$ such that $f_{11}\alpha \neq 0$, a contradiction. We conclude that $A_2 = pA_2$. Similarly, it follows that $A_3 = pA_3$.

3) By Theorem II.6.14 we have to show that all possible maps of the form $\mu f_{11}\alpha$ where μ has domain M_1 and α has codomain A_1 lie in the appropriate Reg. The possibilities are depicted in the following diagram.

By hypothesis we have $\text{Hom}(A_2, A_1) = 0$, and $\text{Hom}(M_1, M_3) = 0$ which settles affirmatively five of the nine possible combinations $\mu f_{11} \alpha$. But also $\text{Hom}(A_3, A_1) = 0$ and $\text{Hom}(M_1, M_2) = 0$ by our assumptions. This settles three more cases affirmatively, and leaves the single composite $\mu f_{11} \alpha : A_1 \to A_1 \xrightarrow{f_{11}} M_1 \to M_1$ which is in $\text{Reg}(A_1, M_1)$ because $\text{Reg}(A_1, M_1) = \text{Hom}(A_1, M_1)$.

4) Suppose that $f \in \text{Reg}(A, M)$ and $f_{33} \neq 0$. Then $\text{Reg}(A_3, M_3) \neq 0$, and therefore A_3 and M_3 are divisible by Theorem 2.13.

5) By Theorem II.6.14 we have to show that all possible maps of the form $\mu f_{33} \alpha$ where μ has domain M_3 and α has codomain A_3 lie in the appropriate Reg. The possibilities are depicted in the following diagram.

$$
\begin{array}{ccccccc}
A_1 & & & & & & M_1 \\
& \searrow & & & & \nearrow & \\
A_2 & \longrightarrow & A_3 & \xrightarrow{f_{33}} & M_3 & \longrightarrow & M_2 \\
& \nearrow & & & & \searrow & \\
A_3 & & & & & & M_3
\end{array}
$$

We have $\text{Hom}(A_1, A_3) = 0$, $\text{Hom}(A_2, A_3) = 0$, and $\text{Hom}(M_3, M_1) = 0$ which settles affirmatively seven of the nine possible combinations $\mu f_{33} \alpha$. But also the composite $\mu f_{11} \alpha : A_3 \to A_3 \xrightarrow{f_{33}} M_3 \to M_2$ is equal to 0 because M_3 is divisible by assumption. Finally, the composite $\mu f_{11} \alpha : A_3 \to A_3 \xrightarrow{f_{33}} M_3 \to M_3$ is regular because $f_{33} \in \text{Hom}(A_3, M_3) = \text{Reg}(A_3, M_3)$.

6) is the combination of 3) and 5). $\qquad\square$

3 Mixed Groups

We now turn to reduced mixed groups and start with some simple examples. The first example shows that A and M may have fairly general direct summands that are irrelevant for regularity.

Example 3.1. Let A_0 and B_0 be elementary groups, D a divisible group and Z a reduced torsion-free group. Let $A = A_0 \oplus D$ and let $M := B \oplus Z$. Then $\text{Hom}(A, M) = \text{Reg}(A, M)$ because $\text{Hom}(A, M) = \text{Hom}(A, B)$.

The second example is a large example where regularity is due to large numbers of direct summands.

Example 3.2. Let T_1, T_2 be elementary groups, let $P_1 = \prod_{p \in \mathbb{P}} (T_1)_p$ and $P_2 = \prod_{p \in \mathbb{P}} (T_2)_p$. For $i = 1, 2$ set $\mathbb{P}(T_i) := \{p \in \mathbb{P} \mid (T_i)_p \neq 0\}$.

1) $\text{Hom}(P_1, P_2) \cong \prod_{p \in \mathbb{P}} \text{Hom}((T_1)_p, (T_2)_p)$, so $f \in \text{Hom}(P_1, P_2)$ may be identified with $f = \prod_{p \in \mathbb{P}} f_p$ where $f_p \in \text{Hom}((T_1)_p, (T_2)_p)$ and the action is componentwise.

2) $\text{Hom}(P_1, P_2) \cong \prod_{p \in \mathbb{P}(T_1) \cap \mathbb{P}(T_2)}^{\cdot} \text{Hom}((T_1)_p, (T_2)_p)$.

3) $\mathrm{Hom}(P_1, P_2) = \mathrm{Reg}(P_1, P_2)$.

Proof. 1) By homological algebra $\mathrm{Hom}(P_1, P_2) \cong \prod_{p \in \mathbb{P}} \mathrm{Hom}(P_1, (T_2)_p)$. We have a short exact sequence $T_1 \rightarrowtail P_1 \twoheadrightarrow P_1/T_1$ where P_1/T_1 is torsion-free and divisible. Hence, for all $p \in \mathbb{P}$,

$$0 = \mathrm{Hom}(\frac{P_1}{T_1}, (T_2)_p) \rightarrowtail \mathrm{Hom}(P_1, (T_2)_p) \to \mathrm{Hom}(T_1, (T_2)_p) \to \mathrm{Ext}(\frac{P_1}{T_1}, (T_2)_p) = 0$$

is exact, and $\mathrm{Hom}(T_1, (T_2)_p) \cong \prod_{q \in \mathbb{P}} \mathrm{Hom}((T_1)_q, (T_2)_p) = \mathrm{Hom}((T_1)_p, (T_2)_p)$. The claim follows because all maps are natural.

2) follows immediately from 1).

3) Let $f = \prod_{p \in \mathbb{P}} f_p \in \mathrm{Hom}(P_1, P_2)$. For any $x = \prod_{p \in \mathbb{P}} x_p \in P_1$ we have $f(x) = \prod_{p \in \mathbb{P}} f_p(x_p)$ which shows that $\mathrm{Ker}(f) = \prod_{p \in \mathbb{P}} \mathrm{Ker}(f_p)$ and $\mathrm{Im}(f) = \prod_{p \in \mathbb{P}} \mathrm{Im}(f_p)$. We also have decompositions

$$(T_1)_p = \mathrm{Ker}(f_p) \oplus (T_1')_p, \quad (T_2)_p = \mathrm{Im}(f_p) \oplus (T_2')_p$$

and we get

$$P_1 = \prod_{p \in \mathbb{P}} \mathrm{Ker}(f_p) \oplus \prod_{p \in \mathbb{P}} (T_1')_p, \text{ and } P_2 = \prod_{p \in \mathbb{P}} \mathrm{Im}(f_p) \oplus \prod_{p \in \mathbb{P}} (T_2')_p.$$

Thus $\mathrm{Ker}(f)$ and $\mathrm{Im}(f)$ are direct summands and f is regular. $\square$

The following "small" example is a modification of the example [11, Vol.II, page 186, Ex.2].

Example 3.3. For each prime $p \in \mathbb{P}$ let $T_p := \langle a_p \rangle$ be cyclic of order p. Let $T = \bigoplus_{p \in \mathbb{P}} T_p$ and let $P = \prod_{p \in \mathbb{P}} T_p$. Let $a = \prod_{p \in \mathbb{P}} a_p \in P$. Then for every $p \in \mathbb{P}$ and every positive integer n there exists a unique element $\hat{a}_{p,n} \in P$ such that

$$a = p^n \hat{a}_{p,n} + a_p.$$

Let $A = T + \sum_{p,n} \mathbb{Z} \hat{a}_{p,n} \subseteq P$. Then the following hold.

1) $\mathbf{t}(A) = T$, $A/\mathbf{t}(A) \cong \mathbb{Q}$, and A is pure in P.

2) If $f \in \mathrm{End}(A)$ and $\mathrm{Im}(f) \subseteq T$, then $\mathrm{Im}(f)$ is finite; furthermore, $A = \mathrm{Im}(f) \oplus \mathrm{Ker}(f)$ and f is regular.

3) The short sequence with natural maps

$$\mathrm{Hom}(A, T) \rightarrowtail \mathrm{End}(A) \twoheadrightarrow \mathrm{End}(A/T) = \mathbb{Q}$$

is exact.

4) $\mathrm{Reg}(A, A) = \mathrm{End}(A)$.

Proof. 1) It is clear that $\mathbf{t}(A) = T$ because $T \subseteq A \subseteq P$ and $T = \mathbf{t}(P)$. It is also clear that $A/T \cong \mathbb{Q}$ because $a + T = p^n \hat{a}_{p,n} + T$ which says that the generators of A/T are all dependent and A/T is divisible.

2) Let $f = \Pi_{p \in \mathbb{P}} f_p \in \operatorname{End}(A)$ where we may and will assume that f_p is (multiplication by) an element in $\mathbb{Z}/p\mathbb{Z}$ because T_p is cyclic of order p. Suppose that $\operatorname{Im}(f) \subseteq T$. Then in particular $f(a) = \Pi_{p \in \mathbb{P}} f_p a_p \in T$. Hence $f_p = 0$ for almost all p and $\operatorname{Im}(f) \subseteq \bigoplus_{f_p \neq 0} T_p$ which is finite. Moreover, it is easily seen that

- actually $\operatorname{Im}(f) = \bigoplus_{f_p \neq 0} T_p$,

- $P = \operatorname{Im}(f) \oplus \prod_{f_p = 0} T_p$,

- $A = \operatorname{Im}(f) \oplus [A \cap \prod_{f_p = 0} T_p]$,

- and $\operatorname{Ker}(f) = A \cap \prod_{f_p = 0} T_p$,

- hence $A = \operatorname{Im}(f) \oplus \operatorname{Ker}(f)$.

3) Let $m \in \mathbb{Z}$ and identify m with multiplication by m in A. Then $\operatorname{Ker}(m) = A[m]$ and $\operatorname{Im}(m) = mA$. It is easily seen that $A[m] = T[m] = P[m] = \bigoplus_{p|m} T_p$, $\prod_{p \nmid m} T_p = mP$, and $P = T[m] \oplus mP$. Since $T[m] \subseteq A$ and A is pure in P, it follows that $A = A[m] \oplus (A \cap mP) = A[m] \oplus mA = \operatorname{Ker}(m) \oplus \operatorname{Im}(m)$. This shows that multiplication by m on A is a regular endomorphism. Let $g \in \operatorname{End}(A)$ be such that $mgm = m$. Let $\bar{g}$ denote the endomorphism A/T induced by g and note that multiplication by m in A/T is an automorphism. Hence $mgm = m$ implies that $m\bar{g} = 1_{A/T}$ and shows that g induces multiplication by $1/m$ on $\operatorname{End}(A/T) \cong \mathbb{Q}$. It is now clear that the natural map $\operatorname{End}(A) \to \operatorname{End}(A/T)$ is surjective and the remaining exactness checks being routine, the claim is established.

4) This was shown in greater generality by Glaz-Wickless ([13, Theorem 4.1]) but we provide an independent proof. Let $f \in \operatorname{End}(A)$. Then the induced endomorphism $\bar{f} \in \operatorname{End}(A/T) \cong \mathbb{Q}$ is regular. Hence there is $g \in \operatorname{End}(A)$ such that $\bar{f}\bar{g}\bar{f} = \bar{f}$ and hence $fgf - f \in \operatorname{Hom}(A, T)$ which by 2) is regular. By Lemma II.4.1 with $M = A$, the endomorphism f itself is regular. $\square$

Proposition 3.5 generalizes an idea used in Example 3.3. We first consider induced maps in general.

Lemma 3.4. *Let A and M be arbitrary groups with maximal torsion subgroups $T_A := \mathbf{t}(A)$ and $T_M := \mathbf{t}(M)$.*

1) *There is a well-defined homomorphism*

$$\overline{} : \operatorname{Hom}(A, M) \to \operatorname{Hom}(A/T_A, M/T_M)$$

given by $\bar{f}(x + T_A) = f(x) + T_M$ where $f \in \operatorname{Hom}(A, M)$ and $x \in A$. Furthermore, $\operatorname{Ker}(\overline{}) = \operatorname{Hom}(A, T_M) \subseteq \operatorname{Hom}(A, M)$.

2) *There is a well-defined homomorphism*

$$\overline{} : \mathrm{Hom}(M, A) \to \mathrm{Hom}(M/T_M, A/T_A)$$

given by $\overline{g}(x + T_M) = g(x) + T_A$ *where* $g \in \mathrm{Hom}(M, A)$ *and* $x \in M$. *Further-more,* $\mathrm{Ker}(\overline{}) = \mathrm{Hom}(M, T_A) \subseteq \mathrm{Hom}(M, A)$.

3) *Let* $\phi_A : A \to A/T_A$ *and* $\phi_T : M \to M/T_M$ *be the natural epimorphisms. Then the maps* $\overline{f}$ *and* $\overline{g}$ *defined in 1) and 2) are the unique mappings satis-fying*

$$\overline{f}\phi_A = \phi_M f, \quad \overline{g}\phi_M = \phi_A g.$$

Proof. 1) The map $\overline{}$ is well-defined since for every $f \in \mathrm{Hom}(A, M)$ it is true that $f(T_A) \subseteq f(T_M)$. Clearly $\overline{}$ is a homomorphism with kernel $\mathrm{Hom}(A, T_M) = \{f \in \mathrm{Hom}(A, M) \mid f(A) \subseteq T_M\}$.

2) is identical with 1) except for the changed roles of A and M and 3) is just a rephrasing to the definition of $\overline{}$. □

Proposition 3.5. *Let* A *be a group such that* $pA_p = 0$, $T_A := \mathbf{t}(A) = \bigoplus_{p \in \mathbb{P}} A_p \subseteq A \subseteq P := \prod_{p \in \mathbb{P}} A_p$, *and* A *is pure in* P. *Let* M *be an arbitrary group with maximal torsion subgroups* $T_M := \mathbf{t}(M)$. *Let* $\overline{} : \mathrm{Hom}(A, M) \to \mathrm{Hom}(A/T_A, M/T_M)$ *be the natural map.*

1) *The quotient* A/T_A *is divisible, and* $\mathrm{Hom}(A/T_A, M/T_M)$ *is a torsion-free divisible group and so a* $\mathbb{Q}$*-vector space.*

2) *Multiplication on* A *by* $m \in \mathbb{N}$ *is a regular element of* $\mathrm{End}(A)$ *and the image of* $\overline{} : \mathrm{End}(A) \to \mathrm{End}(A/T_A)$ *is divisible as an abelian group.*

3) *The image* $\overline{\mathrm{Hom}(A, M)}$ *in the* $\mathbb{Q}$*-vector space* $\mathrm{Hom}(A/T_A, M/T_M)$ *is a* $\mathbb{Q}$*-subspace.*

4) *If* f *is regular in* $\mathrm{Hom}(A, M)$, *then* $\overline{f}$ *is regular in* $\overline{\mathrm{Hom}(A, M)}$.

5) *If* $\overline{f}$ *is regular in* $\overline{\mathrm{Hom}(A, M)}$ *and* $\mathrm{Hom}(A, T_M)$ *is regular, then* f *is regular.*

Proof. 1) It is easy to see that P/T_A is divisible and hence A/T_A is divisible because A is pure in P. Since A/T_A is divisible, therefore $\mathrm{End}(A/T_A)$ is divisi-ble and so the right $\mathrm{End}(A/T_A)$-module $\mathrm{Hom}(A/T_A, M/T_M)$ is also divisible and torsion-free because M/T_M is torsion-free.

2) Let $m \in \mathbb{N}$ be given. Then $A = \left(\bigoplus_{p \mid m} T_p \right) \oplus \left(A \cap \prod_{p \nmid m} T_p \right)$ and the summand $A \cap \prod_{p \nmid m} T_p$ is uniquely m-divisible because A is pure in P and $\prod_{p \nmid m} T_p$ is uniquely divisible in P. Define f_m by stipulating that

$$f_m = \begin{cases} 0 & \text{on } \bigoplus_{p \mid m} T_p \\ \frac{1}{m} & \text{on } A \cap \prod_{p \nmid m} T_p \end{cases}.$$

Then $m f_m m = m$ showing that m is regular and the quasi-inverse f_m induces multiplication by $1/m$ in A/T_A.

3) By 2) for every $m \in \mathbb{N}$, the ring $\mathrm{End}(A)$ contains an element f_m that induces multiplication by $1/m$ in $\mathrm{End}(A/T_A)$. In fact, having $f_m \in \mathrm{End}(A)$ and $f \in \mathrm{Hom}(A, M)$, we obtain $\overline{ff_m} = \frac{1}{m}\overline{f} \in \mathrm{Hom}(A, M)$ showing that $\overline{\mathrm{Hom}(A, M)}$ is divisible.

4) Suppose that $fgf = f$ for some $g \in \mathrm{Hom}(M, A)$. Then

$$\begin{aligned} fgf = f \;\Rightarrow\; & \phi_M fgf = \phi_M f \Rightarrow \overline{f}\phi_A gf = \overline{f}\phi_A \\ \Rightarrow\; & \overline{f}\overline{g}\phi_M f = \overline{f}\phi_A \Rightarrow \overline{f}\overline{g}\overline{f}\phi_A = \overline{f}\phi_A \Rightarrow \overline{f}\overline{g}\overline{f} = \overline{f}, \end{aligned}$$

the last implication being valid because ϕ_A is surjective.

5) Suppose now that $\overline{f}\overline{g}\overline{f} = \overline{f}$ for $g \in \mathrm{Hom}(M, A)$. Then $(fgf - f)(A) \subseteq T_M$, i.e., $fgf - f \in \mathrm{Hom}(A, T_M)$ and so regular by hypothesis. By Lemma IV.4.1 with $M = A$, the homomorphism f itself is regular. $\qquad\square$

If $f \in \mathrm{Reg}(A, M)$, then pf must be regular for every prime p. This fact alone has far-reaching consequences.

Theorem 3.6. *Let A and M be reduced abelian groups and assume that $\mathrm{Reg}(A, M) \neq 0$. Then $\mathbb{P}(A, M) := \{p \in \mathbb{P} \mid \mathrm{Im}(f)[p] \neq 0\}$ is a non-void set of primes such that*

$$\forall p \in \mathbb{P}(A, M), \; A = pA \oplus A[p], \quad A[p] \neq 0,$$

and

$$\forall p \in \mathbb{P}(A, M), \; M = M[p] \oplus M', \quad M[p] \neq 0.$$

Proof. Let $0 \neq f \in \mathrm{Reg}(A, M)$. Then there are decompositions (Proposition 2.9)

$$A = \mathrm{Ker}(f) \oplus A_1, \quad M = \mathrm{Im}(f) \oplus M_1, \quad \forall p \in \mathbb{P} : \mathrm{Im}(f) = p\,\mathrm{Im}(f) \oplus \mathrm{Im}(f)[p],$$

where $A_1 \cong \mathrm{Im}(f)$. Then $\mathbb{P}(A, M) \neq \emptyset$ as otherwise $\mathrm{Im}(f) = p\,\mathrm{Im}(f)$ would be p-divisible for all primes p and hence divisible and so, A being reduced, we would have $\mathrm{Im}(f) = 0$. So far we only used that $\mathbb{Z}f$ is regular. In order to profitably apply Lemma 2.12 we need to find suitable maps α and μ.

Let $p \in \mathbb{P}(A, M)$ and write $\mathrm{Ker}(f) = K_1^p \oplus K_2^p$ such that K_1^p is an elementary p-group and K_2^p has no direct summand of order p. Such decompositions exist by Lemma I.3.1. We have $\mathrm{Im}(f) = p\,\mathrm{Im}(f) \oplus \mathrm{Im}(f)[p]$ where $p\,\mathrm{Im}(f)$ is p-divisible and hence $(p\,\mathrm{Im}(f))_p = 0$. As $A_1 \cong \mathrm{Im}(f)$ we have correspondingly that $A_1 = pA_1 \oplus A_1[p]$ where pA_1 is p-divisible and $(pA_1)_p = 0$. We now have

$$A = K_1^p \oplus K_2^p \oplus A_1[p] \oplus pA_1 \tag{9}$$

where K_1^p and $A_1[p]$ are elementary p-groups and pA_1 contains no p-primary elements. We will show next that $K_2^p = pK_2^p$, i.e., K_2^p is p-divisible. In fact, consider the composite map $\alpha : K_2^p \to K_2^p/pK_2^p \to A_1[p]$. By choice of p we have that $A_1[p] \neq 0$ and if K_2^p/pK_2^p were nonzero also, there must exist a nonzero map α whose kernel would have to be a summand of K_2^p with complementary summand a nonzero elementary p-group. This would contradict the choice of K_2^p, and we conclude that indeed K_2^p is p-divisible. This in turn implies that $K_2^p[p] = 0$ because p-divisible p-groups are divisible and A is reduced. Hence $A_p = A[p] = K_1^p \oplus A_1[p]$ and $pA = pK_2^p \oplus p(pA_1) = K_2^p \oplus pA_1$. Substituting in (9) we get that $A = pA \oplus A[p]$.

Suppose $p \in \mathbb{P}(A, M)$ and write $M_1 = M_{11}^p \oplus M_{12}^p$ such that M_{11}^p is an elementary p-group and M_{12}^p has no direct summand of order p. Such decompositions exist by Lemma I.3.1. We now have

$$M = p \operatorname{Im}(f) \oplus \operatorname{Im}(f)[p] \oplus M_{11}^p \oplus M_{12}^p, \tag{10}$$

where $\operatorname{Im}(f)[p]$ and M_{11}^p are elementary p-groups, $(p \operatorname{Im}(f))[p] = 0$ and M_{12}^p has no direct summand of order p. Suppose that $M_{12}^p[p] \neq 0$. Then there is a nonzero map $\mu : \operatorname{Im}(f) \overset{\text{proj}}{\to} \operatorname{Im}(f)[p] \to M_{12}^p$ and it follows from Lemma 2.12 that $\operatorname{Im}(\mu) \subseteq^\oplus M_{12}^p$. But $\operatorname{Im}(\mu)$ is a nonzero p-elementary group and cannot be a summand by choice of the decomposition $M_1 = M_{11}^p \oplus M_{12}^p$. We conclude that $(M_{12}^p)_p = 0$. Therefore $M[p] = \operatorname{Im}(f)[p] \oplus M_{11}^p$ and substituting in (10) we obtain that $M = M[p] \oplus M'$ where $M' = p \operatorname{Im}(f) \oplus M_{12}^p$. $\qquad\square$

Theorem 3.7. *Let A and M be reduced abelian groups, $S := \operatorname{End}(M)$, $T := \operatorname{End}(A)$, and let $0 \neq f \in \operatorname{Hom}(A, M)$. If SfT is regular if and only if the following conditions* 1) *and* 2) *hold.*

1) $A = \operatorname{Ker}(f) \oplus A_1$, *and* $M = \operatorname{Im}(f) \oplus M_1$ *for some groups* $A_1 \subseteq A$ *and* $M_1 \subseteq M$.

2) *Using the decompositions in* 1) *we have the identifications*

$$f = \begin{bmatrix} 0 & f_{21} \\ 0 & 0 \end{bmatrix}, \mu = \begin{bmatrix} \mu_{11} & \mu_{21} \\ \mu_{12} & \mu_{22} \end{bmatrix} \in S, \alpha = \begin{bmatrix} \alpha_{11} & \alpha_{21} \\ \alpha_{12} & \alpha_{22} \end{bmatrix} \in T, \tag{11}$$

where

$$\mu_{11} : \operatorname{Im}(f) \to \operatorname{Im}(f), \quad \mu_{21} : M_1 \to \operatorname{Im}(f),$$
$$\mu_{12} : \operatorname{Im}(f) \to M_1, \qquad \mu_{22} : M_1 \to M_1,$$

and

$$\alpha_{11} : \operatorname{Ker}(f) \to \operatorname{Ker}(f), \quad \alpha_{21} : A_1 \to \operatorname{Ker}(f),$$
$$\alpha_{12} : \operatorname{Ker}(f) \to A_1, \qquad \alpha_{22} : A_1 \to A_1.$$

With these notations,

i) $\mu_{11} f_{21} \alpha_{12} \in \operatorname{Reg}(\operatorname{Ker}(f), \operatorname{Im}(f))$,

ii) $\mu_{11} f_{21} \alpha_{22} \in \operatorname{Reg}(A_1, \operatorname{Im}(f))$,

iii) $\mu_{12} f_{21} \alpha_{12} \in \operatorname{Reg}(\operatorname{Ker}(f), M_1)$,

iv) $\mu_{12} f_{21} \alpha_{22} \in \operatorname{Reg}(A_1, M_1)$.

Proof. 1) is simply Corollary II.1.3, and 2) is a special case of Theorem II.6.14. $\qquad\square$

Theorem 3.7 by far does not determine $\mathrm{Reg}(A, M)$ for reduced mixed groups because $\mathrm{Reg}(\mathrm{Ker}(f), \mathrm{Im}(f))$, $\mathrm{Reg}(A_1, \mathrm{Im}(f))$, $\mathrm{Reg}(\mathrm{Ker}(f), M_1)$, and $\mathrm{Reg}(A_1, M_1)$ are not known. For one thing, it is quite unclear what $\mathrm{Ker}(f)$ and M_1 might be, however $\mathrm{Im}(f) \cong A_1$ can be described quite well (Lemma 2.7, Proposition 3.10) which however does not mean that $\mathrm{Reg}(A_1, \mathrm{Im}(f)) \cong \mathrm{Reg}(\mathrm{End}(\mathrm{Im}(f)))$ is known. Before we discuss further the determination of $\mathrm{Reg}(\mathrm{End}(\mathrm{Im}(f)))$ we summarize what can be said at this point.

Theorem 3.8. *Let A and M be reduced abelian groups, $S := \mathrm{End}(M)$, $T := \mathrm{End}(A)$, and let $0 \neq f \in \mathrm{Hom}(A, M)$. If SfT is regular, then the following statements hold.*

1) $A = \mathrm{Ker}(f) \oplus A_1$, *and* $M = \mathrm{Im}(f) \oplus M_1$ *for some groups* $A_1 \subseteq A$ *and* $M_1 \subseteq M$.

2) $\forall p \in \mathbb{P} : \mathrm{Im}(f) = p\,\mathrm{Im}(f) \oplus \mathrm{Im}(f)[p]$.

3) $\mathbb{P}(f) := \{p \in \mathbb{P} : (\mathrm{Im}(f))\,[p] \neq 0\} \neq \emptyset$.

4) $f_{21} : A_1 \ni x \mapsto f(x) \in \mathrm{Im}(f)$ *is in* $\mathrm{Reg}(A_1, \mathrm{Im}(f))$.

5) *For every* $\mu_{11} \in \mathrm{Hom}(\mathrm{Im}(f), M_1)$ *it is true that* $\mathrm{Im}(\mu_{11}) \subseteq^{\oplus} M_1$ *and* $\mathrm{Ker}(\mu_{11}) \subseteq^{\oplus} \mathrm{Im}(f)$. *In particular, for* $p \in \mathbb{P}(f)$, *it is true that* $(M_1)_p = M_1[p]$.

6) *For every* $\alpha_{12} \in \mathrm{Hom}(\mathrm{Ker}(f), \mathrm{Im}(f))$ *it is true that* $\mathrm{Im}(\alpha_{12}) \subseteq^{\oplus} \mathrm{Im}(f)$ *and* $\mathrm{Ker}(\alpha_{12}) \subseteq^{\oplus} \mathrm{Ker}(f)$. *In particular, decomposing* $\mathrm{Ker}(f)$ *as* $\mathrm{Ker}(f) = K_1^p \oplus K_2^p$ *such that* K_1^p *is p-elementary and* K_2^p *has no direct summands of order p, it is true for* $p \in \mathbb{P}(f)$ *that* $pK_2 = K_2$.

7) *For every* $\lambda \in \mathrm{Hom}(\mathrm{Ker}(f), M_1)$ *it is true that* $\mathrm{Im}(\lambda) \subseteq^{\oplus} M_1$ *and* $\mathrm{Ker}(\lambda) \subseteq^{\oplus} \mathrm{Ker}(f)$.

Proof. Suppose that SfT is nonzero and regular.

1) and 2) are established in Corollary 2.9.

3) follows from 2) because $\mathbb{P}(f) = \emptyset$ would mean that $\mathrm{Im}(f) = p\,\mathrm{Im}(f)$ for every prime p, i.e., $\mathrm{Im}(f)$ would be a divisible subgroup of a reduced group and hence $\mathrm{Im}(f) = 0$, contrary to hypothesis.

Using the decomposition in 1) we have the identifications

$$f = \begin{bmatrix} 0 & f_{21} \\ 0 & 0 \end{bmatrix}, \quad \mu = \begin{bmatrix} \mu_{11} & \mu_{21} \\ \mu_{12} & \mu_{22} \end{bmatrix} \in S, \quad \alpha = \begin{bmatrix} \alpha_{11} & \alpha_{21} \\ \alpha_{12} & \alpha_{22} \end{bmatrix} \in T, \qquad (12)$$

where

$$\mu_{11} : \mathrm{Im}(f) \to \mathrm{Im}(f), \quad \mu_{21} : M_1 \to \mathrm{Im}(f),$$
$$\mu_{12} : \mathrm{Im}(f) \to M_1, \quad \mu_{22} : M_1 \to M_1,$$

and

$$\alpha_{11} : \mathrm{Ker}(f) \to \mathrm{Ker}(f), \quad \alpha_{21} : A_1 \to \mathrm{Ker}(f),$$
$$\alpha_{12} : \mathrm{Ker}(f) \to A_1, \quad \alpha_{22} : A_1 \to A_1.$$

We now apply Theorem II.6.14 and Lemma 2.12 to get 4), 5), 6) and 7). $\qquad\square$

Lemma 2.7 suggests investigating the consequences of the clearly very restrictive equations $A = pA \oplus A[p]$.

We will elucidate the structure of groups A satisfying $A = pA \oplus A[p]$.

Proposition 3.9. *Let $\mathbb{P}'$ be a set of primes and A be an abelian group such that, for every prime $p \in \mathbb{P}'$, there is a decomposition*

$$A = pA \oplus A_1^p, \quad A_1^p \subseteq A[p].$$

Then $A/D \cong G$ where D is the maximal $\mathbb{P}'$-divisible subgroup of A and G is a group with $\bigoplus_{p \in \mathbb{P}'} A_1^p \subseteq G \subseteq \prod_{p \in \mathbb{P}'} A_1^p$ such that $G/\bigoplus_{p \in \mathbb{P}'} A_1^p$ is $\mathbb{P}'$-divisible.

Proof. Let $p \in \mathbb{P}'$. The decomposition $A = pA \oplus A_1^p$ with $pA_1^p = 0$ implies $pA = p(pA) \oplus pA_1^p = p(pA)$, i.e., pA is p-divisible. Let $\pi_p : A \to A_1^p$ be the projection with kernel pA and let $\pi : A \to P := \prod_{p \in \mathbb{P}'} A_1^p$ be the canonical map induced by the π_p. Then π is the identity on $\bigoplus_{p \in \mathbb{P}'} A_1^p = \mathbf{t}(P)$. Secondly, for every $p \in \mathbb{P}'$, $A/A_1^p \cong pA$ is p-divisible and maps onto $\pi(A)/\mathbf{t}(P)$ which shows that the latter is $\mathbb{P}'$-divisible. It remains to show that $\mathrm{Ker}(\pi) = \bigcap_{p \in \mathbb{P}} pA$ is $\mathbb{P}'$-divisible. Let $x \in \mathrm{Ker}(\pi)$. We will show that for any prime $p \in \mathbb{P}'$, there is $y \in \mathrm{Ker}(\pi)$ such that $x = py$. As $x \in \bigcap_{q \in \mathbb{P}'} qA \subseteq pA$ there is $y' \in A$ such that $x = py'$. Write $y' = x' + y$, where $x' \in A_1^p$ and $y \in pA$. Then $x = py$. Let q be a prime of $\mathbb{P}'$ different from p. We will see that $y \in qA$ so that actually $y \in \bigcap_{q \in \mathbb{P}} qA = \mathrm{Ker}(\pi)$. We have integers u, v such that $1 = up + vq$ and hence $y = upy + vqy = ux + vqy \in qA$ where it is used that $x \in qA$. This establishes the embedding. $\qquad\square$

The following corollaries are immediate.

Corollary 3.10. *Suppose that A is a reduced abelian group such that, for every prime $p \in \mathbb{P}$, there is a decomposition*

$$A = pA \oplus A[p].$$

Then $A \cong G$ where G is a group with $\bigoplus_{p \in \mathbb{P}} A[p] \subseteq G \subseteq \prod_{p \in \mathbb{P}} A[p]$ such that $G/\bigoplus_{p \in \mathbb{P}} A[p]$ is divisible.

Corollary 3.11. *Suppose that $f \in \mathrm{Reg}(A, M)$ and M is reduced. Then $\mathrm{Im}(f) \cong G$ where G is a group with*

$$\bigoplus_{p \in \mathbb{P}} \mathrm{Im}(f)[p] \subseteq G \subseteq \prod_{p \in \mathbb{P}} \mathrm{Im}(f)[p]$$

such that $G/\bigoplus_{p \in \mathbb{P}} \mathrm{Im}(f)[p]$ is divisible.

Proof. Lemma 2.7 and Corollary 3.10. $\qquad\square$

Corollary 3.12. *If in Proposition 3.9 the group A is reduced, then necessarily $A_1^p = A[p]$ and hence $A = A[p] \oplus pA$ and D is torsion-free.*

Proof. We have $pA = p(pA)$ and hence $(pA)_p = 0$ because a p-divisible subgroup of p-group is divisible. For the same reason D must be torsion-free. $\qquad\square$

Theorem 3.6 and Proposition 3.9 suggest considering the following classes of groups.

Definition 3.13. For each $p \in \mathbb{P}$, let T_p be a zero or nonzero elementary p-group, set $P := \prod_{p \in \mathbb{P}} T_p$, $T := \bigoplus_{p \in \mathbb{P}} T_p = \mathbf{t}(P)$, and let $\mathbb{P}(T) := \{p \in \mathbb{P} \mid T_p \neq 0\}$. A group A is a $\mathcal{G}^*(T)$-**group** if $T \subseteq A \subseteq P$ such that A/T is (torsion-free) and $\mathbb{P}(T)$-divisible. Note that $A/T = 0$, i.e., $A = T$, is allowed. A $\mathcal{G}^*(T)$-group A will be called **slim** if T_p is cyclic or zero for every $p \in \mathbb{P}$. The symbol $\mathcal{G}^*(T)$ also stands for the class of all $\mathcal{G}^*(T)$-groups, so that $A \in \mathcal{G}^*(T)$ if and only if A is a $\mathcal{G}^*(T)$-group. A group is a $\mathcal{G}^*$-group if it is a $\mathcal{G}^*(T)$-group for some (elementary) group T. Let $\pi_p : P \to T_p$ denote the projections. The restriction of π_p to a subgroup of P will also be denoted by π_p.

The definition of $\mathcal{G}$-group of Glaz-Wickless (see [13]) differs from our definition of $\mathcal{G}^*$-group in requiring that A/T be divisible and not just $\mathbb{P}(T)$-divisible.

We will first get some insight into the homomorphisms between $\mathcal{G}^*$-groups.

Lemma 3.14. *Let $A \in \mathcal{G}^*(T_1)$ and $M \in \mathcal{G}^*(T_2)$. Then there is a natural map*

$$\sigma : \mathrm{Hom}(A, M) \to \prod_{p \in \mathbb{P}} \mathrm{Hom}((T_1)_p, (T_2)_p)$$

with $\mathrm{Ker}(\sigma) \cong \mathrm{Hom}(A/T_1, M)$. The mapping σ is an embedding if $A/T_1 = p(A/T_1)$ whenever $(T_2)_p \neq 0$. This is the case if A/T_1 is divisible or if $\mathbb{P}(T_2) \subseteq \mathbb{P}(T_1)$.

Proof. The map σ is the composite of the restriction map

$$\mathrm{Hom}(A, M) \to \mathrm{Hom}(T_1, M) = \mathrm{Hom}(T_1, T_2)$$

with the natural isomorphism

$$\mathrm{Hom}(T_1, T_2) \to \prod_{p \in \mathbb{P}} \mathrm{Hom}((T_1)_p, (T_2)_p).$$

Let $f \in \mathrm{Hom}(A, M)$ and let $\phi_1 : A \to A/T_1$ be the natural epimorphism. Assume that $\sigma(f) = 0$. Then $f(T_1) = 0$ and hence $f = f_1 \phi_1$ for some $f_1 \in \mathrm{Hom}(A/T_1, M)$. Conversely, if $f \in \mathrm{Hom}(A, M)$ and $f = f_1 \phi_1$ for some $f_1 \in \mathrm{Hom}(A/T_1, M)$, then $f(T_1) = 0$ and so $\sigma(f) = 0$.

If A/T_1 is divisible, then $\mathrm{Hom}(A/T_1, M) = 0$ since M is reduced. More precisely, M contains no subgroup that is p-divisible for every $p \in \mathbb{P}(T_2)$. On the other hand, by definition of $\mathcal{G}^*(T_1)$, we have $A/T_1 = p(A/T_1)$ for every prime $p \in \mathbb{P}(T_1)$. Consequently, if $\mathbb{P}(T_2) \subseteq \mathbb{P}(T_1)$, then $\mathrm{Hom}(A/T_1, M) = 0$. $\square$

Recall that a group G is **bounded** if there exists a nonzero integer n such that $nG = 0$. Suppose that G is a torsion group with primary decomposition $G = \bigoplus_{p \in \mathbb{P}} G_p$. If G is bounded, then $G_p = 0$ for all but finitely p. Suppose that G_p is finite for every p. Then G is bounded if and only if it is finite. For example, a subgroup of a slim $\mathcal{G}^*$-group is bounded if and only if it is finite.

We begin with some basic observations.

Lemma 3.15. *Let A be a $\mathcal{G}^*(T)$-group. Then the following statements hold.*

1) *A is $\mathbb{P}(T)$-pure in P. More strongly, if $p \in \mathbb{P}(T)$, $x \in P$, and $px \in A$, then $x \in A$.*

2) *If $F \subseteq \mathbb{P}(T)$ is a finite set of primes, then*

$$A = \bigoplus_{p \in F} T_p \oplus \left[A \cap \prod_{p \notin F} T_p \right].$$

Further,

$$\mathbf{t}\left(A \cap \prod_{p \notin F} T_p \right) = \bigoplus_{p \notin F} T_p \text{ and } \frac{\left[A \cap \prod_{p \notin F} T_p \right]}{\bigoplus_{p \notin F} T_p} \cong \frac{A}{\mathbf{t}(A)},$$

the latter group being torsion-free and $\mathbb{P}(T)$-divisible. Furthermore,

$$\forall p \in F, \ p(A \cap \prod_{p \notin F} T_p) = A \cap \prod_{p \notin F} T_p.$$

3) *If B is a bounded subgroup of A, then there is a finite set of primes $F \subseteq \mathbb{P}(T)$, such that $B \subseteq \bigoplus_{p \in F} T_p$ and $B \subseteq^{\oplus} A$.*

4) *Every endomorphism f of A extends uniquely to an endomorphism $\hat{f}$ of P and hence is of the form $f = \prod_{p \in \mathbb{P}} f_p$ where $f_p \in \mathrm{End}(T_p)$.*

5) *Let m also stand for multiplication by m on A and assume that m is a $\mathbb{P}(T)$-number. Then m is regular.*

Proof. 1) Let $x \in P$, $p \in \mathbb{P}(T)$ and $px \in A$. Then $px + T \in A/T$ and as A/T is $\mathbb{P}(T)$-divisible there is $a \in A$ such that $px + T = p(a + T)$. Hence $x - a \in T \subseteq A$ and so $x \in A$.

2) Clearly $P = \left[\bigoplus_{p \in F} T_p \right] \oplus P'$ where $P' := \prod_{p \notin F} T_p$. Since $\bigoplus_{p \in F} T_p \subseteq A$, the decomposition of A follows by intersecting $P = \left[\bigoplus_{p \in F} T_p \right] \oplus P'$ with A. Furthermore, $\mathbf{t}(A \cap \prod_{p \notin F} T_p) = \bigoplus_{p \notin F} T_p$ and hence

$$\frac{A \cap \prod_{p \notin F} T_p}{\bigoplus_{p \notin F} T_p} \cong \frac{(\bigoplus_{p \in F} T_p) \oplus (A \cap \prod_{p \notin F} T_p)}{(\bigoplus_{p \in F} T_p) \oplus (\bigoplus_{p \notin F} T_p)} = \frac{A}{\mathbf{t}(A)}$$

is torsion-free and $\mathbb{P}(T)$-divisible.

Let $p \in F$. Then clearly $P' = pP'$ and $A \cap P'$ is p-pure in P' because $A = T_p \oplus A \cap P'$ and A is p-pure in P, hence $A \cap P' = (A \cap P') \cap pP' = p(A \cap P')$.

3) The bounded subgroup B has a primary decomposition $B = \bigoplus_{p \in \mathbb{P}} B_p$ with $0 \neq B_p \subseteq T_p$ for p in some finite set of primes F. Since $\bigoplus_{p \in F} T_p$ is semisimple and contains B, it follows that B is a direct summand of $\bigoplus_{p \in F} T_p$ which in turn is a summand of A, so B is a summand of A.

4) The maximal torsion subgroup T of P is fully invariant in P. Therefore there is a well-defined restriction map $\mathrm{End}(P) \to \mathrm{End}(T) \cong \prod_{p \in \mathbb{P}} \mathrm{End}(T_p)$ that is an isomorphism. Hence elements of $\mathrm{End}(P) \cong \mathrm{End}(T)$ are of the form $f = \Pi_{p \in \mathbb{P}} f_p$.

Let $f \in \mathrm{End}(A)$. First restrict f to $\mathbf{t}(A) = T = \mathbf{t}(P)$ and then extend to an endomorphism of P that restricts to f on A.

5) Let $F := \{p \in \mathbb{P} \mid p \text{ divides } m\}$. Then $F \subseteq \mathbb{P}(T)$. Clearly $\mathrm{Ker}(m) = \bigoplus_{p \in F} T_p$ and by 1)

$$A = \left[\bigoplus_{p \in F} T_p \right] \oplus \left[A \cap \prod_{p \notin F} T_p \right].$$

We will show that $mA = A \cap \prod_{p \notin F} T_p$. Then $A = \mathrm{Ker}(m) \oplus \mathrm{Im}(m)$ which shows that m is regular. It is clear that $mA \subseteq A \cap \prod_{p \notin F} T_p$. Conversely, by 2), $A \cap \prod_{p \notin F} T_p = m(A \cap \prod_{p \notin F} T_p) \subseteq mA$. $\square$

The following theorem shows that it is quite common to have nonzero regular bimodules in $\mathrm{Hom}(A, M)$ for $\mathcal{G}^*$-groups A and M.

Theorem 3.16. *Suppose that $A \in \mathcal{G}^*(T_1)$ and $M \in \mathcal{G}^*(T_2)$. Then $\mathbf{t}(\mathrm{Hom}(A, M))$ is a regular bi-submodule of $\mathrm{Hom}(A, M)$.*

Proof. Note that $f \in \mathbf{t}(\mathrm{Hom}(A, M))$ if and only if $\mathrm{Im}(f)$ is bounded. Suppose that $f \in \mathbf{t}(\mathrm{Hom}(A, M))$, $\mu \in \mathrm{End}(M)$, and $\alpha \in \mathrm{End}(A)$. Then $(\mu f \alpha)(A) \subseteq (\mu f)(A) = \mu(f(A))$ which is bounded. Hence $\mathbf{t}(\mathrm{Hom}(A, M))$ is a bi-submodule of $\mathrm{Hom}(A, M)$.

Let $f \in \mathbf{t}(\mathrm{Hom}(A, M))$, and let $\mathbb{P}_1 := \{p \in \mathbb{P} \mid (\mathrm{Im}(f))_p \neq 0\}$. Then $\mathbb{P}_1$ is finite and $\mathrm{Im}(f) \subseteq^{\oplus} \bigoplus_{p \in \mathbb{P}_1} (T_2)_p \subseteq^{\oplus} M$. Set $P_1 := \prod_{p \in \mathbb{P}} (T_1)_p$. Then $P_1 = \left[\bigoplus_{p \in \mathbb{P}_1} (T_1)_p \right] \oplus \left[\prod_{p \notin \mathbb{P}_1} (T_1)_p \right]$ and $A = \left[\bigoplus_{p \in \mathbb{P}_1} (T_1)_p \right] \oplus A \cap \left[\prod_{p \notin \mathbb{P}_1} (T_1)_p \right]$. Note that $\mathbb{P}_1 \subseteq \mathbb{P}(T_1)$. It now follows from the fact that $f(\bigoplus_{p \notin \mathbb{P}_1} T_p) = 0$ and the fact that $\left(A \cap \prod_{p \notin \mathbb{P}_1} (T_1)_p \right) / \sum_{p \notin \mathbb{P}_1} (T_1)_p$ is $\mathbb{P}(T_1)$-divisible, hence $\mathbb{P}_1$-divisible, that $A \cap \left[\prod_{p \notin \mathbb{P}_1} (T_1)_p \right] \subseteq \mathrm{Ker}(f)$. Hence $\mathrm{Ker}(f) = \left[\mathrm{Ker}(f) \cap \bigoplus_{p \in \mathbb{P}_1} (T_1)_p \right] \oplus A \cap \left[\prod_{p \notin \mathbb{P}_1} (T_1)_p \right]$ and since $\mathrm{Ker}(f) \cap \left[\bigoplus_{p \in \mathbb{P}_1} (T_1)_p \right]$ is a direct summand of the elementary group $\bigoplus_{p \in \mathbb{P}_1} (T_1)_p$, it follows that $\mathrm{Ker}(f) \subseteq^{\oplus} A$. This shows that f is regular. $\square$

The following example shows that a $\mathcal{G}^*(T)$-group may have epimorphic images of order p if $p \notin \mathbb{P}(T)$. This means that the map σ in Lemma 3.14 need not be an embedding without proper assumptions.

Example 3.17. Let $\mathbb{P}_1$ be a proper infinite subset of $\mathbb{P}$. For every $p \in \mathbb{P}_1$ let T_p be a group of order p, let $T := \bigoplus_{p \in \mathbb{P}_1} T_p$, and let $P := \prod_{p \in \mathbb{P}_1} T_p$. Let $A \subseteq P$ be the group given by $A/T \cong Q := \{\frac{m}{n} \in \mathbb{Q} \mid n \text{ is a } \mathbb{P}_1\text{-number}\}$. Then A/T is $\mathbb{P}_1$-divisible, so $A \in \mathcal{G}^*(T)$ but, for any $p \notin \mathbb{P}_1$ we obtain a mapping $A \to A/T \cong Q \to Q/pQ \cong \mathbb{Z}(p)$.

Combining earlier results we obtain a characterization of $\mathcal{G}^*$-groups.

Theorem 3.18. *Let A be an abelian group. Then A is a $\mathcal{G}^*(T)$-group for $T := \bigoplus_{p \in \mathbb{P}} A_p$ if and only if A is reduced and*

$$\forall p \in \mathbb{P}(T), \ A = pA \oplus A[p].$$

Proof. Suppose that $A \in \mathcal{G}^*$. Then A is reduced and by Lemma 3.15.1 with $F = \{p\} \subseteq \mathbb{P}(T)$ we have $A = T_p \oplus [A \cap \prod_{q \neq p} T_q]$ where $T_p = A[p]$ and $A \cap \prod_{q \neq p} T_q$ is p-divisible. Hence $pA = p[A \cap \prod_{q \neq p} T_q] = [A \cap \prod_{q \neq p} T_q]$.

The converse is settled in Corollary 3.10.   □

The following example shows that for $A \in \mathcal{G}^*(T)$, the equality $A = A[p] \oplus pA$ need not hold for all p.

Example 3.19. Let $T = \bigoplus_{p \in \mathbb{P}} T_p$ be the elementary abelian group with $T_2 = 0$, $a_2 = 0$, and, for primes $p \geq 3$, let $T_p = \langle a_p \rangle$ be cyclic of order p. Let $a = \prod_{p \in \mathbb{P}} a_p \in P := \prod_{p \in \mathbb{P}} T_p$. Let A be the subgroup of P given by $A/T = \mathbb{Z}_{(2)}(a+T)$ where $\mathbb{Z}_{(2)}$ is $\mathbb{Z}$ localized at the prime ideal $(2) = 2\mathbb{Z}$. Then $A \in \mathcal{G}^*(T)$ but $A \neq A[2] \oplus 2A$.

Proof. It is clear that $A \in \mathcal{G}^*(T)$ and $A_2 = 0$. Hence the equality $A = A[2] \oplus 2A$ is simply $A = 2A$. By way of contradiction assume that there is $x \in A$ such that $2x = a$. Since $x \in A$, there is $m/n \in \mathbb{Z}_{(2)}$ such that $x + T = \frac{m}{n}(a + T)$, or equivalently, $nx - ma \in T$ where n is odd and $\gcd(m,n) = 1$. It follows that $(n - 2m)x \in T$ with $n - 2m \neq 0$, and this implies that $x \in T$, and so $a = 2x \in T$, a contradiction.   □

The following example shows that there are nice groups in $\mathcal{G}^*(T)$ yet there are non-regular homomorphisms between these groups. The example demonstrates that it will not be easy to determine those pairs of mixed groups A and M such that $\mathrm{Hom}(A, M)$ is regular.

Example 3.20. For each $p \in \mathbb{P}$, let $T_p = \langle a_p \rangle$ be cyclic of order p, let $T := \bigoplus_{p \in \mathbb{P}} T_p$, $P := \prod_{p \in \mathbb{P}} T_p$, $a = \prod_{p \in \mathbb{P}} a_p$, and let A be the purification of $T + \mathbb{Z}a$ in P (see Lemma I.3.5). Split $\mathbb{P}$ into a disjoint union $\mathbb{P} = \mathbb{P}_1 \cup \mathbb{P}_2$ such that both parts $\mathbb{P}_1$ and $\mathbb{P}_2$ are infinite. Let $b := \prod_{p \in \mathbb{P}_2} a_p$ and let M be the purification of $T + \mathbb{Z}b$ in P. Define

$$f = \prod_{p \in \mathbb{P}} f_p \text{ such that } f_p = \begin{cases} 0 & \text{if } p \in \mathbb{P}_1 \\ 1 & \text{if } p \in \mathbb{P}_2 \end{cases}.$$

Then $f \in \mathrm{Hom}(A, M)$, $M = \bigoplus_{p \in \mathbb{P}_1} T_p \oplus \mathrm{Im}(f)$, $\mathrm{Ker}(f) = \bigoplus_{p \in \mathbb{P}_1} T_p$, and $\mathrm{Ker}(f)$ is not a direct summand of A.

Proof. To establish that $f \in \mathrm{Hom}(A, M)$ requires only to check that $f(a) \in M$ because obviously $f(T) \subseteq T \subseteq M$ and $f(T + \mathbb{Z}a) \subseteq M$ will imply that $f((T + \mathbb{Z}a)_*^P) \subseteq (f(T) + \mathbb{Z}f(a))_*^P \subseteq M$. But clearly, $f(a) = \prod_{p \in \mathbb{P}} f_p a_p = \prod_{p \in \mathbb{P}_2} a_p = b \in M$.

It follows from the definition of f that $\mathrm{Ker}(f) = \bigoplus_{p \in \mathbb{P}_1} T_p$ and $\mathrm{Im}(f) = M \cap \prod_{p \in \mathbb{P}_2} T_p$ so $M = \bigoplus_{p \in \mathbb{P}_1} T_p \oplus \mathrm{Im}(f)$. It remains to show that $\mathrm{Ker}(f)$ is not a summand of A. By way of contradiction assume that $A = \mathrm{Ker}(f) \oplus B$ for some subgroup B of A. We have $P = P_1 \oplus P_2$ where $P_1 = \prod_{p \in \mathbb{P}_1} T_p$ and $P_2 = \prod_{p \in \mathbb{P}_2} T_p$

and it is immediately seen that P_1 is the maximal $\mathbb{P}_2$-divisible subgroup of P and P_2 is the maximal $\mathbb{P}_1$-divisible subgroup of P. Now $\mathbf{t}(B) = \bigoplus_{p \in \mathbb{P}_2} T_p$ is $\mathbb{P}_1$-divisible and $B/\mathbf{t}(B) \cong (\bigoplus_{p \in \mathbb{P}_1} T_p \oplus B)/(\bigoplus_{p \in \mathbb{P}_1} T_p \oplus \mathbf{t}(B)) = A/\mathbf{t}(A) \cong \mathbb{Q}$, hence by Lemma I.3.6, B is $\mathbb{P}_1$-divisible and therefore $B \subseteq P_2$. It is now clear that $a \notin \mathrm{Ker}(f) \oplus B$ and we have arrived at the desired contradiction. $\qquad\square$

4 Regularity in Endomorphism Rings of Mixed Groups

It has been seen in Theorem 3.8 that $\mathrm{Reg}(A, A) = \mathrm{Reg}(\mathrm{End}(A))$ comes into play when studying $\mathrm{Hom}(A, M)$. Unfortunately, while there are several papers on the subject ([29], [12], [30], [13], [24]), it is not even known which mixed abelian groups have regular endomorphism rings. We include some of the more accessible results but the subject is wide open, and even more so the computation of $\mathrm{Reg}(A, M)$ for mixed groups A, M.

We begin with an immediate but attractive corollary of Proposition 2.9.

Corollary 4.1. *For any abelian group $\mathbb{Z}1_A \subseteq \mathrm{End}(A)$ is regular if and only if $A = nA \oplus A[n]$ for every $n \in \mathbb{Z}$.*

The neatest result on regular endomorphism rings is due to Glaz and Wickless for which we provide a new proof.

Theorem 4.2. ([13, Theorem 4.1]) *Let A be a slim $\mathcal{G}^*(T)$-group such that A/T has finite rank and is divisible. Then $\mathrm{End}(A)$ is regular.*

Proof. Let $f \in \mathrm{End}(A)$. Then $f = \prod_{p \in \mathbb{P}} f_p$ where $f_p \in \mathbb{Z}/p\mathbb{Z}$ and we may choose $f_p = 0$ if $T_p = 0$. Let $\mathbb{P}_1 := \{p \in \mathbb{P} \mid f_p = 0\}$ and let $\mathbb{P}_2 := \{p \in \mathbb{P} \mid f_p \neq 0\}$. Then clearly

$$\mathrm{Ker}(f) \subseteq \prod_{p \in \mathbb{P}_1} T_p, \quad \text{and} \quad \mathrm{Im}(f) \subseteq \prod_{p \in \mathbb{P}_2} T_p.$$

Hence

$$\mathrm{Ker}(f) \cap \mathrm{Im}(f) = 0, \quad \text{also} \quad f(T) = \bigoplus_{p \in \mathbb{P}_2} T_p. \qquad (13)$$

We also have a short exact sequence

$$\frac{\mathrm{Ker}(f) + T}{T} \rightarrowtail \frac{A}{T} \twoheadrightarrow \frac{f(A)}{f(T)}.$$

Observe next that

$$\begin{aligned} f(T) \subseteq T \cap f(A) &= [(\oplus_{p \in \mathbb{P}_1} T_p) \oplus (\oplus_{p \in \mathbb{P}_2} T_p)] \cap f(A) \\ &\subseteq [\mathrm{Ker}(f) \oplus f(T)] \cap f(A) = f(T), \end{aligned}$$

and so $T \cap f(A) = f(T)$. Therefore

$$\frac{f(A)}{f(T)} = \frac{f(A)}{T \cap f(A)} \cong \frac{f(A) + T}{T} \subseteq \frac{A}{T}.$$

It follows that $f(A)/f(T)$ is torsion-free and therefore $\frac{\mathrm{Ker}(f)+T}{T}$ is pure in A/T and therefore divisible. We conclude that the short exact sequence splits and

$$\dim \frac{A}{T} = \dim \frac{\mathrm{Ker}(f) + T}{T} + \dim \frac{f(A)}{f(T)}. \tag{14}$$

We also have

$$\frac{A}{T} \supseteq \frac{\mathrm{Ker}(f) + T}{T} \oplus \frac{f(A) + T}{T}. \tag{15}$$

That the sum is direct can be seen as follows.

$$
\begin{aligned}
(\mathrm{Ker}(f) + T) \cap (f(T) + T) &= [\mathrm{Ker}(f) \cap (f(T) + T)] + T \\
&= [\mathrm{Ker}(f) \cap (f(T) + \oplus_{p \in \mathbb{P}_2} T_p)] + T \\
&= (\mathrm{Ker}(f) \cap f(T)) + \oplus_{p \in \mathbb{P}_2} T_p + T = T.
\end{aligned}
$$

The dimension equation (14) together with (15) now implies that $A = \mathrm{Ker}(f) + f(A) + T = \mathrm{Ker}(f) \oplus \mathrm{Im}(f)$ and we have established that f is regular. □

Remark. The proof of Proposition 4.5 also shows that for a $\mathcal{G}^*$-group A the ideal $I = \{f \in \mathrm{End}(A) \mid f(A) \text{ is torsion}\}$ need not be regular.

Inspection of the verification of Example 3.3 and [13, Theorem 3.5]) suggest the following theorem.

Theorem 4.3. *Suppose that A is a $\mathcal{G}^*$-group with the property that every torsion endomorphic image of A is bounded. Then if $\overline{\mathrm{End}(A)}$ is regular, so is $\mathrm{End}(A)$. More generally,*

$$\mathrm{Reg}(\mathrm{End}(A)) = \mathrm{Reg}(\overline{\mathrm{End}(A)}).$$

Proof. Let $f \in \mathrm{Reg}(\mathrm{End}(A))$. Then $fgf = f$ for some $g \in \mathrm{End}(A)$ and hence $\overline{f}\overline{g}\overline{f} = \overline{f}$, i.e., $\overline{f}$ is regular in $\overline{\mathrm{End}(A)}$. Let $\overline{g_1}, \overline{g_2} \in \overline{\mathrm{End}(A)}$. Then $\overline{g_1}\overline{f}\overline{g_2} \in \mathrm{Reg}(\overline{\mathrm{End}(A)})$ and by the established result $\overline{g_1}\overline{f}\overline{g_2}$ is regular. So $\overline{f} \in \mathrm{Reg}(\overline{\mathrm{End}(A)})$.

Conversely, let $f \in \mathrm{End}(A)$ and suppose that $\overline{f} \in \mathrm{Reg}(\overline{\mathrm{End}(A)})$. Then there is $g \in \mathrm{End}(A)$ such that $\overline{f}\overline{g}\overline{f} = \overline{f}$. It follows that $f - fgf \in \mathrm{End}(A)$ is such that $(f - fgf)(A) \subseteq T$. By hypothesis $(f - fgf)(A)$ is bounded and by Theorem 3.16 it follows that $f - fgf$ is regular in $\mathrm{End}(A)$. By Lemma II.4.1 f is regular. Suppose that $\overline{f} \in \mathrm{Reg}(\overline{\mathrm{End}(A)})$ and $g_1, g_2 \in \mathrm{End}(A)$. Then $\overline{g_1 f g_2} = \overline{g_1}\overline{f}\overline{g_2} \in \mathrm{Reg}(\overline{\mathrm{End}(A)})$ and therefore $g_1 f g_2$ is regular by the previous result. Hence $f \in \mathrm{Reg}(\mathrm{End}(A))$. □

We conclude this section with two interesting results on regularity in endomorphism rings.

Theorem 4.4. *Let A be a reduced abelian group and assume that $0 \neq \mathrm{Reg}(A, A)$. Then there exists a non-void set of primes $\mathbb{P}'$ such that*

$$\forall p \in \mathbb{P}', \ A = pA \oplus A[p], \quad A[p] \neq 0.$$

Conversely, suppose that A is a group (reduced or not) and $\mathbb{P}'$ a non-void set of primes such that

$$\forall p \in \mathbb{P}', \ A = pA \oplus A[p], \ and \ A_p = A[p] \neq 0.$$

Then $\mathrm{Reg}(A, A) \neq 0$.

Proof. The first part is the special case $A = M$ of Theorem 3.6.

Now suppose that A is a group and $\mathbb{P}'$ a non-void set of primes such that

$$\forall p \in \mathbb{P}', \ A = pA \oplus A[p], \ and \ A_p = A[p] \neq 0.$$

By induction we obtain for a finite set of primes $p_1, \ldots, p_k \in \mathbb{P}'$ that

$$A = A[p_1] \oplus \cdots \oplus A[p_k] \oplus p_1 \cdots p_k A. \tag{16}$$

Recall that n is a $\mathbb{P}'$-number if the prime factors of n all belong to $\mathbb{P}'$. Let

$$I := \{\xi \in \mathrm{End}(A) \mid n\xi = 0 \text{ for some nonzero } \mathbb{P}'\text{-number } n\}.$$

We claim that I is a nonzero regular ideal of $\mathrm{End}(A)$. First of all $I \neq 0$ because $\mathbb{P}' \neq \emptyset$ and for $p \in \mathbb{P}'$ the projector ξ of A onto $A[p]$ along pA belongs to I. It is more or less evident that I is an ideal in $\mathrm{End}(A)$. It is left to show that every $\xi \in I$ is regular. Let $n \neq 0$ be a $\mathbb{P}'$-number such that $n\xi = 0$ and let $p_1, \ldots, p_k$ be the prime factors of n. Then we have (16) and $n(\mathrm{Im}(\xi)) = 0$. Hence $\mathrm{Im}(\xi) \subseteq A[p_1] \oplus \cdots \oplus A[p_k]$ and the latter being elementary (= semisimple), $\mathrm{Im}(\xi)$ is a direct summand of it and in turn of A. The kernel $\mathrm{Ker}(\xi)$ must contain $p_1 \cdots p_k A$ and therefore it follows from (16) that

$$\mathrm{Ker}(\xi) = (\mathrm{Ker}(\xi) \cap (A[p_1] \oplus \cdots \oplus A[p_k])) \oplus p_1 \cdots p_k A.$$

As $\mathrm{Ker}(\xi) \cap (A[p_1] \oplus \cdots \oplus A[p_k])$ is a direct summand of $A[p_1] \oplus \cdots \oplus A[p_k]$ it follows from (16) that $\mathrm{Ker}(\xi)$ is a direct summand of A. Thus ξ is regular. $\qquad \square$

The next proposition shows that there are also obstructions to regularity.

Proposition 4.5. *Let A be a $\mathcal{G}^*(T)$-group with the property that every torsion direct summand of A is bounded and let B be an elementary group such that $B_p \neq 0$ whenever $T_p \neq 0$. Let $G = A \oplus B$. Then $\mathrm{End}(G)$ is not regular and $\mathrm{Reg}(G, G)$ is a proper ideal of $\mathrm{End}(G)$.*

Proof. The hypotheses provide for the existence of an endomorphism f of G that is 0 on A and maps B onto an unbounded torsion summand of A. Then $f(G)$ is not a direct summand of A and therefore not a summand of G. So f is not regular. Finally, by Theorem 4.4, $\mathrm{Reg}(G, G) \neq 0$. $\qquad \square$

Chapter IX

Regularity in Categories

1 Regularity in Preadditive Categories

The definition of a regular map makes perfectly good sense in any category. To wit, let $\mathcal{C}$ be a category. We write $A \in \mathcal{C}$ if A is an object of $\mathcal{C}$ and by $\mathcal{C}(A, B)$ we denote the set of morphisms in $\mathcal{C}$ for objects A and B in $\mathcal{C}$. Then $f \in \mathcal{C}(A, B)$ is **regular** if and only if there exists $g \in \mathcal{C}(B, A)$ such that $fgf = f$. To do more we need more special categories. A **preadditive category** is a category $\mathcal{C}$ in which the morphism sets are abelian groups with the property that composition of morphisms distributes over addition. It is also postulated that the category contain a null object 0 such that $\mathcal{C}(0, B) = \{0\}$ and $\mathcal{C}(A, 0) = \{0\}$. This assures that $\mathcal{C}(A) := \mathcal{C}(A, A)$ is a ring with $1_A \in \mathcal{C}(A)$ for any object A in the preadditive category $\mathcal{C}$ and $\mathcal{C}(A, M)$ is a $\mathcal{C}(M)$-$\mathcal{C}(A)$-bimodule.

Surprisingly $\operatorname{Reg}(A, M)$ exists in preadditive categories as the largest $\mathcal{C}(M)$-$\mathcal{C}(A)$-submodule of $\mathcal{C}(A, M)$ all of whose morphisms are regular. In fact, inspection of the proofs in the case of module categories reveal that they consist of computation with R-homomorphisms that are equally valid for morphisms in a preadditive category.

In the following it is always assumed that $\mathcal{C}$ is a *preadditive category*.

Lemma 1.1. *Let* $f \in \mathcal{C}(A, M)$ *and* $g \in \mathcal{C}(M, A)$. *If* $f - fgf$ *is regular, then* f *is regular.*

Proof. The proof of Lemma II.4.1 applies literally. $\qquad \square$

Definition 1.2. Let $S := \mathcal{C}(M)$ and $T := \mathcal{C}(A)$. For $f \in \mathcal{C}(A, M)$ let

$$\operatorname{Reg}(A, M) := \{f \in \mathcal{C}(A, M) \mid SfT \text{ is regular}\},$$

where SfT is the S-T-submodule of $\mathcal{C}(A, M)$ generated by f.

A subset X of $\mathcal{C}(A, M)$ is **regular** if every element of X is regular.

Theorem 1.3. $\operatorname{Reg}(A, M)$ *is the largest regular S-T-submodule of $\mathcal{C}(A, M)$.*

Proof. The proof of Theorem II.4.3 applies.   □

2   Preadditive Categories

An object A in a preadditive category $\mathcal{C}$ is a **biproduct** of the objects $A_1, \ldots, A_n \in \mathcal{C}$ if there exist morphisms,

$$\iota_i : A_i \to A, \quad \text{and} \quad \pi_i : A \to A_i,$$

called **structural maps**, such that

$$\pi_i \iota_i = 1_{A_i}, \quad \pi_j \iota_i = 0 \text{ for } i \neq j, \quad \text{and} \quad \iota_1 \pi_1 + \cdots + \iota_n \pi_n = 1_A.$$

If this is the case we write $A \in_\mathcal{C} A_1 \oplus \cdots \oplus A_n$, we say the A is a **biproduct in $\mathcal{C}$** of the objects $A_1, \ldots, A_n$, and think of $A_1 \oplus \cdots \oplus A_n$ as the collection of all data sets (A, ι_i, π_i).

We write $A \in_\mathcal{C} A_1 \oplus \cdots \oplus A_n$ instead of $A = A_1 \oplus \cdots \oplus A_n$ as is often done because there will later be situations (see Section 3, particularly Example 3.12) where we have to deal with true set-theoretic equality as well as with categorical notions.

Let $\mathcal{C}$ be a preadditive category. If $A \in_\mathcal{C} A_1 \oplus \cdots \oplus A_n$ and B is isomorphic with A, then $B \in_\mathcal{C} A_1 \oplus \cdots \oplus A_n$ also. If $A, B \in_\mathcal{C} A_1 \oplus \cdots \oplus A_n$, then A and B are isomorphic. If $A \in_\mathcal{C} A_1 \oplus \cdots \oplus A_n$ and $A_i \cong_\mathcal{C} A_i'$, then $A \in_\mathcal{C} A_1' \oplus \cdots \oplus A_n'$. Further note that the apparent ordering of the summands is introduced by the choice of labels and is not intrinsic to the definition.

A biproduct has the universal properties of both products and coproducts and is unique up to isomorphism in the category.

We recall here the definitions of kernels, cokernels, images and coimages of morphisms in a category.

Definition 2.1. Let $\mathcal{C}$ be a preadditive category and $f \in \mathcal{C}(A, M)$.

1) A **kernel** of f is a morphism $k : K \to A$ such that $fk = 0$ and whenever $\chi : X \to A$ is such that $f\chi = 0$, then there is a unique morphism $\lambda : X \to K$ such that $k\lambda = \chi$. We write $k \in \operatorname{Ker}(f)$ if k is a kernel of f.

2) A **cokernel** of f is a morphism $c : M \to C$ such that $cf = 0$ and whenever $\chi : M \to X$ is such that $\chi f = 0$, then there is a unique morphism $\lambda : C \to X$ such that $\lambda c = \chi$. We write $c \in \operatorname{Coker}(f)$ if c is a cokernel of f.

3) An **image** of f is a kernel of a cokernel of f; we write $\operatorname{Im}(f) = \operatorname{Ker}(\operatorname{Coker}(f))$.

4) A **coimage** of f is the cokernel of a kernel of f, $\operatorname{Coim}(f) = \operatorname{Coker}(\operatorname{Ker}(f))$.

All these concepts are unique up to isomorphism. Recall that a morphism e is **epic** if it can be canceled on the right: $fe = ge \Rightarrow f = g$, and a morphism m is **monic** if it can be canceled on the left: $mf = mg \Rightarrow f = g$.

Example 2.2.

1) Kernels and images are monic.

2) Cokernels and coimages are epic.

Proof. Let $k \in \text{Ker}(f : A \to M)$. Suppose that $kk_1 = kk_2$ for $k_1, k_2 \in \mathcal{C}(B, A)$. Then $f(kk_1) = 0$ and $f(kk_2) = 0$. By definition of kernel the morphism $kk_1 = kk_2$ factors uniquely through k, so $k_1 = k_2$. Images are monic because they are kernels. The epic property of cokernels and coimages is dual. $\qquad\square$

Example 2.3. Let $A \in_{\mathcal{C}} A_1 \oplus A_2$ with structural maps

$$\iota_i : A_i \to A, \quad \text{and} \quad \pi_i : A \to A_i.$$

Let $i, j \in \{1, 2\}$, $i \neq j$. Then

1) $0 : 0 \to A_i$ is a kernel of $\iota_i : A_i \to A$,

2) $0 : A_i \to 0$ is a cokernel of $\pi_i : A \to A_i$,

3) $\iota_i : A_i \to A$ is a kernel of $\pi_j : A \to A_j$,

4) $\pi_i : A \to A_i$ is a cokernel of $\iota_j : A_i \to A$,

5) $\iota_i : A_i \to A$ is an image of $\iota_i : A_i \to A$,

6) $1_{A_i} : A_i \to A_i$ is an image of $\pi_i : A \to A_i$,

7) $1_{A_i} : A_i \to A_i$ is a coimage of $\iota_i : A_i \to A$,

8) $\pi_i : A \to A_i$ is a coimage of $\pi_i : A_i \to A$.

We summarize the results in a table.

	ι_1	π_1	ι_2	π_2
Ker	0	ι_2	0	ι_1
Coker	π_2	0	π_1	0
Im	ι_1	1_{A_1}	ι_2	1_{A_2}
Coim	1_{A_1}	π_1	1_{A_2}	π_2

Proof. 1) and 2) follow from the fact that kernels are monic and cokernels are epic.

3) We have that $\pi_j \iota_i = 0$. Suppose that $\chi : X \to A$ is such that $\pi_j \chi = 0$. Then $\pi_i \chi : X \to A_i$ and $\iota_i \pi_i \chi = (1 - \iota_j \pi_j)\chi = \chi$. So χ factors through A_i. Suppose that $\phi : X \to A_i$ is such that $\iota_i \phi = \chi$. Then $\phi = \pi_i \iota_i \phi = \pi_i \chi$, so ϕ is unique.

4) The proof is dual to 3).

5) $\text{Im}(\iota_i) = \text{Ker}(\text{Coker}(\iota_i) = \text{Ker}(\pi_j) = \iota_i$.

6) $\text{Im}(\pi_i) = \text{Ker}(\text{Coker}(\pi_i) = \text{Ker}(0 : A_i \to 0) = \iota_i$.

7) $\text{Coim}(\iota_i) = \text{Coker}(\text{Ker}(\iota_i)) = \text{Coker}(0 : 0 \to A_i) = 1_{A_i}$.

8) $\text{Coim}(\pi_i) = \text{Coker}(\text{Ker}(\pi_i)) = \text{Coker}(\iota_j) = \pi_i$. $\qquad\square$

Example 2.4. Let $e \in \mathcal{C}(A, A)$ be an idempotent and suppose that e and $1 - e$ have kernels and cokernels. Then

$$\mathrm{Im}(e) = \mathrm{Ker}(1 - e), \quad \text{and} \quad \mathrm{Im}(1 - e) = \mathrm{Ker}(e).$$

Proof. Let $\gamma : A \to C$ be a cokernel of e. Since $(1 - e)e = 0$, there is $\psi : C \to A$ such that

$$1 - e = \psi\gamma. \tag{1}$$

Let $\kappa : K \to A$ be a kernel of $1 - e$. We will show that $\kappa \in \mathrm{Ker}(\gamma) = \mathrm{Ker}(\mathrm{Coker}(e)) = \mathrm{Im}(e)$ thereby establishing in particular that $\mathrm{Ker}(\gamma)$ exists. We have $\gamma\kappa = \gamma(e + (1 - e))\kappa = \gamma e\kappa + \gamma(1 - e)\kappa = 0$. Suppose that $\chi : X \to A$ is such that $\gamma\chi = 0$. Then $(1 - e)\chi \overset{(1)}{=} \psi\gamma\chi = 0$ and since $\kappa \in \mathrm{Ker}(1 - e)$, the map χ factors uniquely through κ showing that $\kappa \in \mathrm{Ker}(\gamma) = \mathrm{Ker}(\mathrm{Coker}(e)) = \mathrm{Im}(e)$.

We now assume that $\lambda : L \to A \in \mathrm{Im}(e) = \mathrm{Ker}(\gamma)$ and show that $\lambda \in \mathrm{Ker}(1-e)$. We have $\gamma\lambda = 0$ and from $\gamma e = 0$ it follows that there is $\xi : A \to L$ such that $e = \lambda\xi$. Suppose that $\chi : X \to A$ is such that $(1 - e)\chi = 0$. As $\kappa \in \mathrm{Ker}(1 - e)$ there is $\zeta : X \to K$ such that $\chi = \kappa\zeta$. It follows that $\chi = e\chi = \lambda(\xi\chi)$. So χ factors through λ and that this is uniquely so follows from the fact that λ is a kernel and hence monic. We have shown that $\lambda \in \mathrm{Ker}(1 - e)$. $\qquad\square$

An **additive category** is a preadditive category that has biproducts for any finite set of objects.

Given an idempotent e in a category, it is not automatic that it can be factored as $e = \iota\pi$, $\pi\iota = 1$ and that it produces a decomposition.

Definition 2.5. Let A be an object in a category $\mathcal{C}$ and $e = e^2 \in \mathcal{C}(A, A)$. Then the **idempotent e splits in $\mathcal{C}$** if there exists an object M and mappings $\iota \in \mathcal{C}(M, A)$, $\pi \in \mathcal{C}(A, M)$ such that $\iota\pi = e$ and $\pi\iota = 1_M$. We say that **idempotents split in $\mathcal{C}$** if all idempotents in $\mathcal{C}$ are splitting.

We observe next that the splitting of idempotents means that every idempotent determines a direct decomposition.

Lemma 2.6. *Let $\mathcal{C}$ be a preadditive category. Suppose that A is an object of $\mathcal{C}$, and e is an idempotent in $\mathcal{C}(A)$. Then $1_A - e$ is another idempotent of $\mathcal{C}(A)$. Suppose further that there are objects and morphisms*

$$\iota_1 : A_1 \to A, \quad \pi_1 : A \to A_1, \quad \iota_2 : A_2 \to A, \quad \pi_2 : A \to A_2$$

such that

$$\iota_1\pi_1 = e, \quad \pi_1\iota_1 = 1_{A_1}, \quad \iota_2\pi_2 = 1_A - e, \quad \pi_2\iota_2 = 1_{A_2}.$$

Then $A \in_{\mathcal{C}} A_1 \oplus A_2$ with structural maps $\iota_1, \pi_1, \iota_2, \pi_2$. In addition,

$$\iota_1 \in \mathrm{Ker}(1_A - e) = \mathrm{Im}(e), \quad \iota_2 \in \mathrm{Ker}(e) = \mathrm{Im}(1_A - e),$$

$$\pi_1 \in \mathrm{Coker}(1_A - e), \quad \pi_2 \in \mathrm{Coker}(e).$$

Proof. We have $\iota_1\pi_1 + \iota_2\pi_2 = e + 1_A - e = 1_A$, next $\pi_1\iota_2 = 1_{A_1}\pi_1\iota_2 1_{A_2} = \pi_1\iota_1\pi_1\iota_2\pi_2\iota_2 = \pi_1 e(1_A - e)\iota_2 = 0$, and similarly $\pi_2\iota_1 = 0$. This shows that A is a biproduct of A_1 and A_2.

We show next that $\iota_1 \in \mathrm{Ker}(1-e)$. First $(1-e)\iota_1\pi_1 = (1-e)e = 0$ and since π_1 is epic, $(1-e)\iota_1 = 0$. Now suppose that $\chi : X \to A$ is such that $(1-e)\chi = 0$. Then $\chi = 1_A\chi = \iota_1\pi_1\chi + \iota_2\pi_2\chi = \iota_1(\pi_1\chi) + (1-e)\chi = \iota_1(\pi_1\chi)$ and this factorization through ι_1 is unique because ι_1 is monic. By symmetry we have $\iota_2 \in \mathrm{Ker}(e)$. It was shown in Example 2.4 that $\mathrm{Im}(e) = \mathrm{Ker}(1 - e)$, and $\mathrm{Im}(1 - e) = \mathrm{Ker}(e)$.

We show next that $\pi_1 \in \mathrm{Coker}(1 - e)$. First $\iota_1\pi_1(1 - e) = e(1 - e) = 0$ and so $\pi_1(1 - e) = 0$ because ι_1 is monic. Suppose that $\chi : A \to X$ is such that $\chi(1 - e) = 0$. Then $\chi = \chi e = (\chi\iota_1)\pi_1$ so χ factors through π_1 and uniquely so because π_1 is epic. By symmetry $\pi_2 \in \mathrm{Coker}(e)$. $\qquad\square$

If idempotents split in a preadditive category $\mathcal{C}$, then we have the familiar decompositions associated with regular maps.

Theorem 2.7. *Let $\mathcal{C}$ be a preadditive category in which idempotents split and suppose that $f \in \mathcal{C}(A, M)$ is regular and $fgf = f$ for $g \in \mathcal{C}(M, A)$. Then the following statements hold.*

1) *$e := fg \in \mathcal{C}(M, M)$ is an idempotent.*

2) *$d := gf \in \mathcal{C}(A, A)$ is an idempotent.*

3) *There are structural maps $\iota_{A_i} : A_i \to A$, and $\pi_{A_i} : A \to A_i$ such that $d = \iota_{A_1}\pi_{A_1}$, $\pi_{A_1}\iota_{A_1} = 1_{A_1}$, $1_A - d = \iota_{A_2}\pi_{A_2}$, $\pi_{A_2}\iota_{A_2} = 1_{A_2}$, and $A \in_{\mathcal{C}} A_1 \oplus A_2$.*

4) *There are structural maps $\iota_{M_i} : M_i \to M$, and $\pi_{M_i} : M \to M_i$ such that $e = \iota_{M_1}\pi_{M_1}$, $\pi_{M_1}\iota_{M_1} = 1_{M_1}$, $1_M - e = \iota_{M_2}\pi_{M_2}$, $\pi_{M_2}\iota_{M_2} = 1_{M_2}$, and $M \in_{\mathcal{C}} M_1 \oplus M_2$.*

5) *$\pi_{M_1}f\iota_{A_1} : A_1 \to M_1$ is an isomorphism with inverse $\pi_{A_1}g\iota_{M_1}$.*

6) *$\iota_{A_2} \in \mathrm{Ker}(f)$ and $\iota_{M_2} \in \mathrm{Ker}(fg)$.*

7) *$\iota_{M_1} \in \mathrm{Im}(f)$ and $\iota_{A_1} \in \mathrm{Im}(gf)$.*

Proof. 1) and 2). It follows from $fgf = f$ that $e = fg$ and $d = gf$ are idempotents.

3) and 4) As idempotents split in $\mathcal{C}$ we have the indicated structural maps and by Lemma 2.6 we have the claimed biproducts.

5) We compute: $\pi_{A_1}g\iota_{M_1}\pi_{M_1}f\iota_{A_1} = \pi_{A_1}gef\iota_{A_1} = \pi_{A_1}gfgf\iota_{A_1} = \pi_{A_1}gf\iota_{A_1} = \pi_{A_1}\iota_{A_1}\pi_{A_1}\iota_{A_1} = 1_{A_1}1_{A_1} = 1_{A_1}$, and $\pi_{M_1}f\iota_{A_1}\pi_{A_1}g\iota_{M_1} = \pi_{M_1}fdg\iota_{M_1} = \pi_{M_1}fgfg\iota_{M_1} = \pi_{M_1}fg\iota_{M_1} = \pi_{M_1}\iota_{M_1}\pi_{M_1}\iota_{M_1} = 1_{M_1}1_{M_1} = 1_{M_1}$.

6) To show that $\iota_{A_2} \in \mathrm{Ker}(f)$ we first note that $f\iota_{A_2}\pi_{A_2} = f(1_A - d) = f(1_A - gf) = 0$ which implies that $f\iota_{A_2} = 0$ because π_{A_2} is epic. Suppose that $\chi : X \to A$ is such that $f\chi = 0$. Then $\iota_{A_2}(\pi_{A_2}\chi) = (1_A - d))\chi = \chi - gf\chi = \chi$. This shows that χ factors through ι_{A_2}. To show that this is uniquely so, suppose that $\iota_{A_2}\phi = \chi$ for some ϕ. Then $\pi_{A_2}\iota_{A_2}\phi = \pi_{A_2}\chi$, therefore $\phi = \pi_{A_2}\chi$.

To show that $\iota_{M_2} \in \mathrm{Ker}(fg)$ we first note that $fg\iota_{M_2}\pi_{M_2} = fg(1_M - e) = fg - fgfg = 0$. Suppose that $\chi : X \to M$ is such that $fg\chi = 0$. Then $\iota_{M_2}\pi_{M_2}\chi = (1 - e)\chi = \chi - fg\chi = \chi$, so χ factors through ι_{M_2}. To show that this is uniquely so, suppose that $\iota_{M_2}\phi = \chi$. Then $\phi = \pi_{M_2}\iota_{M_2}\phi = \pi_{M_2}\chi$.

7) We show first that $\pi_{M_2} \in \mathrm{Coker}(f)$. Indeed, $\iota_{M_2}\pi_{M_2}f = (1 - fg)f = 0$, and, ι_{M_2} being monic, $\pi_{M_2}f = 0$. Suppose that $\chi : M \to X$ is such that $\chi f = 0$. Then $\chi = \chi 1_M = \chi\iota_{M_1}\pi_{M_1} + \chi\iota_{M_2}\pi_{M_2} = \chi fg + \chi\iota_{M_2}\pi_{M_2} = (\chi\iota_{M_2})\pi_{M_2}$, thus χ factors through π_{M_2} and because π_{M_2} is epic, uniquely so. We now have $\iota_{M_1} \in \mathrm{Ker}(\pi_{M_2}) = \mathrm{Ker}(\mathrm{Coker}(f)) = \mathrm{Im}(f)$. It follows from Lemma 2.6 that $\iota_{A_1} \in \mathrm{Im}(gf)$ because $gf = d$ is an idempotent. $\qquad\square$

Let $f \in \mathcal{C}(A, M)$. We wish to generalize to categories the result "f is regular if $\mathrm{Ker}(f) \subseteq^{\oplus} A$ and $\mathrm{Im}(f) \subseteq^{\oplus} M$".

Theorem 2.8. *Let $\mathcal{C}$ be a preadditive category and let $f \in \mathcal{C}(A, M)$. Assume that there exists a biproduct*

$$K \underset{\pi_K}{\overset{\iota_K}{\underset{\rightleftarrows}{}}} A \underset{\iota_L}{\overset{\pi_L}{\underset{\rightleftarrows}{}}} L$$

with $\iota_K \in \mathrm{Ker}(f)$ and assume that

$$I \underset{\pi_I}{\overset{\iota_I}{\underset{\rightleftarrows}{}}} M \underset{\iota_J}{\overset{\pi_J}{\underset{\rightleftarrows}{}}} J$$

is a biproduct with $\pi_J \in \mathrm{Coker}(f)$, and $\iota_I \in \mathrm{Ker}(\pi_J) = \mathrm{Im}(f)$.

Then $\pi_I f \iota_L : L \to I$ is both epic and monic. Assume that there is a morphism $h : I \to L$ such that

$$h\,(\pi_I f \iota_L) = 1_L, \quad and \quad (\pi_I f \iota_L)\, h = 1_I.$$

Then $g : M \to A : g := \iota_L h \pi_I$ is such that $fgf = f$.

Proof. We first record two immediate consequences of our assumptions that we will have to use:

$$f\iota_K = 0, \quad \pi_J f = 0. \tag{2}$$

Suppose that $\chi : X \to L$ is such that $\pi_I f \iota_L \chi = 0$. Then $0 = \iota_I \pi_I f \iota_L \chi = (1_M - \iota_J \pi_J)f\iota_L\chi = f\iota_L\chi - \iota_J(\pi_J f)\iota_L\chi = f\iota_L\chi$. Since $f(\iota_L\chi) = 0$ and $\iota_K \in \mathrm{Ker}(f)$, there is $\phi : X \to K$ such that

$$\iota_L\chi = \iota_K\phi. \tag{3}$$

We conclude that $\chi = 1_L\chi = \pi_L\iota_L\chi \overset{(3)}{=} \pi_L\iota_K\phi = 0$ showing that $\pi_I f \iota_L$ is monic.

Now suppose that $\chi : I \to X$ is such that $\chi\pi_I f \iota_L = 0$. Then $0 = \chi\pi_I f \iota_L \pi_L = \chi\pi_I f(1_A - \iota_K\pi_K) = \chi\pi_I f$ using (2). Since $(\chi\pi_I)f = 0$ and $\pi_J \in \mathrm{Coker}(f)$, there is $\psi : J \to X$ such that

$$\chi\pi_I = \psi\pi_J. \tag{4}$$

We conclude that $\chi = \chi 1_I = \chi \pi_I \iota_I \overset{(4)}{=} \psi \pi_J \iota_I = 0$ showing that $\pi_I f \iota_L$ is epic.

By assumption there is $h : I \to L$ such that $h(\pi_I f \iota_L) = 1_L$ and $(\pi_I f \iota_L) h = 1_I$. Let $g : M \to A : g = \iota_L h \pi_I$. Note that

$$\pi_I f g = \pi_I f \iota_L h \pi_I = 1_I \pi_I = \pi_I. \tag{5}$$

Note further that

$$f = 1_M f = (\iota_I \pi_I + \iota_J \pi_J) f \overset{(2)}{=} \iota_I \pi_I f. \tag{6}$$

Then $fgf \overset{(6)}{=} \iota_I \pi_I fgf \overset{(5)}{=} \iota_I \pi_I f \overset{(6)}{=} f$. $\qquad\qquad\square$

Question 2.9. In a module category the isomorphism $\mathrm{Hom}_R(R, M) \cong M$ leads to regularity in modules. Is there an analogue in categories? Suppose there is an object $A \in \mathcal{C}$ that is free on one generator. Then $\mathcal{C}(A, M) \cong M$.

3 The Quasi-Isomorphism Category of Torsion-free Abelian Groups

In group and module theory there is a special interest in decompositions into direct sums of indecomposable subobjects. For convenience we call a direct decomposition an **indecomposable decomposition** if the direct summands are all indecomposable. In general, torsion-free abelian groups are notorious for their essentially different indecomposable decompositions. In this section we demonstrate that the lack of uniqueness can be salvaged at the expense of introducing an equivalence of groups that is weaker than isomorphism. This leads to the so-called "quasi-isomorphism category".

We begin with an example, that may be considered folklore, of a group X that has two essentially different indecomposable decompositions. The group X is a **finite essential extension** of a completely decomposable group A, meaning that A is large in X and X/A is finite. The fact that A is essential in X is equivalent to saying that X is also torsion-free. The group A has a homogeneous direct summand of rank 2 that can be decomposed in different ways and as a consequence the elements of X ("clamps") that tie the summands of A together are replaced by clamps that tie together different summands of A. The group may be depicted as follows. The homogeneous block of rank 2 pictured on the left is decomposed in two different ways which replaces a single 3-pronged clamp by two 2-pronged clamps.

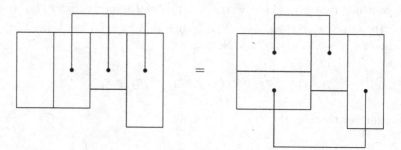

Example 3.1. Let V be a four-dimensional $\mathbb{Q}$-vector space with basis $\{v_1, v_2, v_3, v_4\}$. Then

$$V = \mathbb{Q}v_1 \oplus \mathbb{Q}v_2 \oplus \mathbb{Q}v_3 \oplus \mathbb{Q}v_4.$$

The example will be obtained as an additive subgroup of V. Let $A_0 := \mathbb{Z}[5^{-1}]$, $A_1 := \mathbb{Z}[7^{-1}]$, $A_2 := \mathbb{Z}[11^{-1}]$,

$$A = A_0 v_1 \oplus A_0 v_2 \oplus A_1 v_3 \oplus A_2 v_4, \text{ and } X = A + \mathbb{Z}\frac{1}{2 \cdot 3}(v_2 + 3v_3 + 2v_4).$$

Then

$$
\begin{aligned}
X &= A_0 v_1 \oplus \left((A v_2 \oplus A_1 v_3 \oplus A_2 v_4) + \mathbb{Z}\frac{1}{2 \cdot 3}(v_2 + 3v_3 + 2v_4) \right) \\
&= \left((A_0(2v_1 + v_2) \oplus A_1 v_3) + \mathbb{Z}\frac{1}{2}(2v_1 + v_2 + 3v_3) \right) \\
&\oplus \left((A(3v_1 + v_2) \oplus A_2 v_4) + \mathbb{Z}\frac{1}{3}(3v_1 + v_2 + 2v_4) \right)
\end{aligned}
$$

are indecomposable decompositions with summands of rank 1 and 3 in the first decomposition and two summands of rank 2 in the second decomposition.

Proof. Details can be found in [20, Chapter V, Example 3.1].   □

A very striking and easily stated result on "pathological decomposition" is as follows.

Theorem 3.2 (Corner [6], or [23, page 281]). *Given integers $n \geq k \geq 1$, there exists a torsion-free group X of rank n such that for any partition $n = r_1 + \cdots + r_k$, there is a decomposition of X into a direct sum of k indecomposable subgroups of ranks $r_1, \ldots, r_k$ respectively.*

Let $\mathcal{A}$ denote the category of torsion-free abelian groups of finite rank with the usual homomorphisms $\mathcal{A}(A, M) = \operatorname{Hom}(A, M)$ as morphisms. The torsion-free groups are exactly the additive subgroups of $\mathbb{Q}$-vector spaces. Formally the torsion-free group A is naturally embedded in $\mathbb{Q} \otimes_{\mathbb{Z}} A$, a $\mathbb{Q}$-vector space. This can be ignored if we simply assume that A is an additive subgroup of some $\mathbb{Q}$-vector space V. If so, the subspace of V spanned by A is denoted by $\mathbb{Q}A$ and

$$\mathbb{Q}A = \{ra \mid r \in \mathbb{Q}, a \in A\} (\cong \mathbb{Q} \otimes_{\mathbb{Z}} A).$$

The **rank** of A is $\mathrm{rk}(A) = \dim(\mathbb{Q}A)$.

We insert here a fact that is very helpful and will be used in the following without explicit reference.

Lemma 3.3. *Let $f \in \mathrm{Hom}(A, M)$. Then f has a unique extension $F : \mathbb{Q}A \to \mathbb{Q}M$ that is a linear mapping. Hence any homomorphism $f : A \to M$ may be considered to be the restriction of a linear mapping $F : \mathbb{Q}A \to \mathbb{Q}M$.*

Proof. Let $f \in \mathrm{Hom}(A, M)$ be given. We define $F : \mathbb{Q}A \to \mathbb{Q}M$ as follows. For $x = \frac{1}{n}y \in \mathbb{Q}A$, $y \in A$, let $F(x) = \frac{1}{n}f(y) \in \mathbb{Q}M$. It is an easy exercise to check that F is well-defined (independent of the choice of n and y), linear, unique and restricts to f on A. □

Notational Convention. We will use $A, B, C, M, \ldots$ to denote torsion-free abelian groups in $\mathcal{A}$, letters $f, g, h, \ldots$ for ordinary homomorphisms in $\mathcal{A}$ and capitals $F, G, H, \ldots$ for the linear transformations that restrict to $f, g, h, \ldots$ respectively.

The problem with "pathological decompositions" can be remedied by changing the category as follows. This idea goes back to Bjarni Jonsson ([16], [17]).

Let A and M be groups in $\mathcal{A}$. Recall Lemma 3.3 which can be interpreted to mean that

$$\mathrm{Hom}(A, M) \subseteq \mathrm{Hom}_{\mathbb{Q}}(\mathbb{Q}A, \mathbb{Q}M) = \mathrm{Hom}(\mathbb{Q}A, \mathbb{Q}M),$$

where $\mathrm{Hom}_{\mathbb{Q}}(\mathbb{Q}A, \mathbb{Q}M)$ is the vector space of linear transformations on $\mathbb{Q}A$ to $\mathbb{Q}M$. Let

$$\mathbb{Q}\,\mathrm{Hom}(A, M) = \{rf \mid r \in \mathbb{Q}, f \in \mathrm{Hom}(A, M)\} \subseteq \mathrm{Hom}(\mathbb{Q}A, \mathbb{Q}M)$$

be the subspace of $\mathrm{Hom}(\mathbb{Q}A, \mathbb{Q}M)$ spanned by $\mathrm{Hom}(A, M)$. If A, M, K are three groups in $\mathcal{A}$, $F \in \mathbb{Q}\,\mathrm{Hom}(A, M) \subseteq \mathrm{Hom}(\mathbb{Q}A, \mathbb{Q}M)$, and $G \in \mathbb{Q}\,\mathrm{Hom}(M, K) \subseteq \mathrm{Hom}(\mathbb{Q}M, \mathbb{Q}K)$, then the composite GF of the linear mappings G, F is not only in $\mathrm{Hom}(\mathbb{Q}A, \mathbb{Q}M)$ but in $\mathbb{Q}\,\mathrm{Hom}(A, M)$. Also, if $F, G \in \mathbb{Q}\,\mathrm{Hom}(A, M)$, then $F + G \in \mathbb{Q}\,\mathrm{Hom}(A, M)$. This says that the "category" defined next is indeed a category.

Definition 3.4. Let $\mathbb{Q}\mathcal{A}$ denote the category whose objects are those of $\mathcal{A}$, and whose morphism sets are the abelian groups $\mathbb{Q}\,\mathrm{Hom}(A, M)$ with addition and composition inherited from $\mathrm{Hom}(\mathbb{Q}A, \mathbb{Q}M)$. This is the **quasi-isomorphism category** of torsion-free abelian groups of finite rank.

Proposition 3.5. *The quasi-isomorphism category $\mathbb{Q}\mathcal{A}$ is an additive category.*

Proof. The algebraic laws of the morphisms are automatic because the operations are inherited from vector space laws, each morphism set $\mathbb{Q}\,\mathrm{Hom}(A, M)$ is an abelian group, and the group $\{0\}$ is the zero object of the category. Let $A_1, \ldots A_n$ be in $\mathbb{Q}\mathcal{A}$. Then the ordinary biproduct $A_1 \oplus \cdots \oplus A_n$ in $\mathcal{A}$ is a biproduct in $\mathbb{Q}\mathcal{A}$. □

The following lemma contains the basic properties of the quasi-isomorphism category that are essential for working purposes. The routine verifications are left to the reader.

Lemma 3.6.

1) $\mathbb{Q}\operatorname{Hom}(A, M) = \{F \in \operatorname{Hom}(\mathbb{Q}A, \mathbb{Q}M) \mid \exists\, 0 \neq n \in \mathbb{N}, nF(A) \subseteq M\} = \{\frac{1}{n}f \mid 0 \neq n \in \mathbb{N}, f \in \operatorname{Hom}(A, M)\}$.

2) Let $F, G \in \mathbb{Q}\operatorname{Hom}(A, M)$. Choose $0 \neq n \in \mathbb{N}$ such that $nF \in \operatorname{Hom}(A, M)$ and $nG \in \operatorname{Hom}(A, M)$. Then $F = G$ in $\mathbb{Q}A$ if and only if $nF = nG$ in $\operatorname{Hom}(A, M)$.

3) Let $F, G \in \mathbb{Q}\operatorname{Hom}(A, M)$. Choose $0 \neq n \in \mathbb{N}$ such that $nF \in \operatorname{Hom}(A, M)$ and $0 \neq m \in \mathbb{N}$ such that $mG \in \operatorname{Hom}(A, M)$. Then $F + G = \frac{1}{nm}(nF + mG)$.

4) Let $F \in \mathbb{Q}\operatorname{Hom}(A, M)$ and $G \in \mathbb{Q}\operatorname{Hom}(M, K)$. Choose $0 \neq n \in \mathbb{N}$ such that $f := nF \in \operatorname{Hom}(A, M)$ and $0 \neq m \in \mathbb{N}$ such that $g := mG \in \operatorname{Hom}(M, K)$. Then $GF = \frac{1}{nm}gf$. Hence $GF = 0$ if and only if $gf = 0$.

5) Let $F \in \mathbb{Q}\operatorname{Hom}(A, M)$ and $G \in \mathbb{Q}\operatorname{Hom}(M, K)$. Choose $0 \neq n \in \mathbb{N}$ such that $f := nF \in \operatorname{Hom}(A, M)$ and $0 \neq m \in \mathbb{N}$ such that $g := mG \in \operatorname{Hom}(M, K)$. If $GF = H \in \mathbb{Q}\operatorname{Hom}(A, K)$, then $gf = mnH \in \operatorname{Hom}(A, K)$.

Two groups A, M that are isomorphic in the category $\mathbb{Q}A$ are **quasi-isomorphic** and we write $A \cong_{\mathrm{qu}} M$. All $\mathbb{Q}A$ notions can be expressed in A. It is immediate that isomorphic groups are quasi-isomorphic. The interpretation of isomorphism in $\mathbb{Q}A$ is as follows.

Proposition 3.7. *Let $A, M \in \mathbb{Q}A$. Then the following statements are equivalent.*

1) $A \cong_{\mathrm{qu}} M$.

2) *There exist $0 \neq n \in \mathbb{N}$, $f \in \operatorname{Hom}(A, M)$, $g \in \operatorname{Hom}(M, A)$ such that $fg = n \cdot 1_M$ and $gf = n \cdot 1_A$.*

3) *There exists a monomorphism $f : A \to M$ and $0 \neq n \in \mathbb{N}$ such that $nM \subseteq f(A) \subseteq M$.*

Proof. 1) $\Rightarrow$ 2) Suppose $A \cong_{\mathrm{qu}} M$. Then there are $\frac{1}{k}f \in \mathbb{Q}\operatorname{Hom}(A, M)$, $\frac{1}{m}g \in \mathbb{Q}\operatorname{Hom}(M, A)$ such that $\frac{1}{k}f \cdot \frac{1}{m}g = 1_{\mathbb{Q}M}$ and $\frac{1}{m}g \cdot \frac{1}{k}f = 1_{\mathbb{Q}A}$. Hence $n = km$ will do.

2) $\Rightarrow$ 3) Suppose that $fg = n \cdot 1_M$ and $gf = n \cdot 1_A$, then f is injective and $nM = n \cdot 1_M M = fg(M) \subseteq f(A) \subseteq M$.

3) $\Rightarrow$ 1) Suppose that $f : A \to M$ is a monomorphism and $0 \neq n \in \mathbb{N}$ such that $nM \subseteq f(A) \subseteq M$. Define $g : M \to A : g := f^{-1}n$. Then $\left(\frac{1}{n}f\right)g = 1_M$ and $g\left(\frac{1}{n}f\right) = 1_A$ hence $A \cong_{\mathrm{qu}} M$. □

Example 3.8. The group X in Example 3.1 is quasi-isomorphic to its completely decomposable subgroup A: $X \cong_{\mathrm{qu}} A$.

Proof. The inclusion of A in X is a monomorphism and the index $[X : A]$ is finite, in fact, a short computation shows that $[X : A] = 6$. $\square$

Our next example deals with **rational groups** that are defined to be additive subgroups of $\mathbb{Q}$ that contain $\mathbb{Z}$. Every additive subgroup of $\mathbb{Q}$ is isomorphic with a rational group and it is convenient to have 1 as an element of the group because of Lemma 3.10. We need a standard result from number theory.

Lemma 3.9 (Partial Fraction Decomposition). *Let n be a positive integer and let $n = \prod_p p^{n_p}$ be its prime factorization. Then there exist integers u_p such that*

$$\tfrac{1}{n} = \sum_p \tfrac{u_p}{p^{n_p}}.$$

Rational groups can be obtained by choosing some set of generators. For example, $\mathbb{Z} = \langle 1 \rangle$ and $\mathbb{Q} = \left\langle \frac{1}{p^n} : p \in \mathbb{P}, n \in \mathbb{N}_0 \right\rangle$. The goal is to gain control of the generating sets of rational groups.

Lemma 3.10. *Let A be a rational group.*

1) *If $\frac{m}{n} \in A$ with $\gcd(m, n) = 1$, then $\frac{1}{n} \in A$ and A is generated by its elements of the form $\frac{1}{p^k}$ where $p \in \mathbb{P}$.*

2) *If $\frac{1}{m}, \frac{1}{n} \in A$, then $\frac{1}{\text{lcm}(m,n)} \in A$.*

Proof. 1. Assume that $\frac{m}{n} \in A$ with $\gcd(m, n) = 1$. Choose integers u, v such that $um + vn = 1$. Then $\frac{1}{n} = u\frac{m}{n} + v \in A$. Hence A is generated by its fractions $\frac{1}{n}$. Let $\frac{1}{n}$ be such a generator and consider its prime factorization $n = \prod p^{n_p}$. By Lemma 3.9,

$$\frac{1}{n} = \sum_p \frac{u_p}{p^{n_p}} \in A.$$

This shows that the special fractions $1/p^{n_p} \in A$ generate A.

2. Write $\gcd(m, n) = um + vn$. Then

$$\frac{1}{\text{lcm}(m, n)} = \frac{\gcd(m, n)}{mn} = u\frac{1}{n} + v\frac{1}{m} \in A. \qquad \square$$

We can now clarify quasi-isomorphism for groups of rank 1.

Example 3.11. Two rank-1 groups are quasi-isomorphic if and only if they are isomorphic.

Proof. We can assume without loss of generality that the two quasi-isomorphic groups are rational groups because every rank-1 group is isomorphic to a rational group. Let A and B be rational groups that are quasi-isomorphic. Any homomorphism $A \to B$ is multiplication by a rational number and hence injective. By assumption, there is $r \in \mathbb{Q}$ such that $mB \subseteq rA \subseteq B$ (Proposition 3.7.3). Since $A \cong rA$ we assume without loss of generality that $A = rA$, so that we have

$mB \subseteq A \subseteq B$. We now look at generators of the form $1/p^s$, p a prime. All such generators of A are contained in B, and if $1/p^s \in B$, then $m/p^s \in A$, and if $\gcd(m,p) = 1$, then in fact $1/p^s \in A$. Hence it is only generators $1/p^s$ with p dividing m, that may prevent A from being equal to B. One easily convinces oneself that $1/p^s \in B$ for all s implies that $1/p^s \in A$ for all s. This leaves the case that p divides m and not all $1/p^s$ are in B. In this case there is a largest S such that $1/p^S \in B$. Because $A \subseteq B$ there is a largest T such that $1/p^T \in A$ and necessarily $T \leq S$. Now $p^{S-T}B \cong B$ and the largest generator $1/p^r \in p^{S-T}B$ is $1/p^T$, the same as A, while the other generators have not changed because of Lemma 3.10.1. Hence $mp^{S-T}B \subseteq mB \subseteq A \subseteq p^{S-T}B$. This process can be repeated for other prime factors of m and we end up with $A = m'B$ for some factor m' of m. We arrived at $A = m'B \cong B$ as desired.   □

The following example may serve as motivation and explanation for writing $A \in_\mathcal{C} A_1 \oplus \cdots \oplus A_n$ instead of $A = A_1 \oplus \cdots \oplus A_n$ as is often done.

Example 3.12. Let $A = \mathbb{Z}v_1 \oplus \mathbb{Z}v_2$ in $\mathcal{A}$ which is true set-theoretic equality. Let $M = K = \mathbb{Z}(v_1 + v_2)$. Then $A \in_{\mathbb{Q}\mathcal{A}} M \oplus K$ but $A = M \oplus K$ would not make sense strictly speaking.

Proof. Structural maps for $A \in_{\mathbb{Q}\mathcal{A}} M \oplus K$ are

$$
\begin{aligned}
\iota_M : M \to A &: \quad \iota_M(v_1 + v_2) = v_1, \\
\pi_M : A \to M &: \quad \pi_M(v_1) = v_1 + v_2, \pi_M(v_2) = 0; \\
\iota_K : K \to A &: \quad \iota_K(v_1 + v_2) = v_2, \\
\pi_K : A \to M &: \quad \pi_M(v_1) = 0, \pi_K(v_2) = v_1 + v_2.
\end{aligned}
$$

Of course, it also follows from $A \in_{\mathbb{Q}\mathcal{A}} \mathbb{Z}v_1 \oplus \mathbb{Z}v_2$ and $\mathbb{Z}v_1 \cong_{\mathrm{qu}} \mathbb{Z}(v_1 + v_2) \cong_{\mathrm{qu}} \mathbb{Z}v_2$ that $A \in_{\mathbb{Q}\mathcal{A}} M \oplus K$.   □

We now interpret the meaning of biproduct in $\mathbb{Q}\mathcal{A}$ in the category $\mathcal{A}$. While $M \oplus K$ as such is meaningful in $\mathcal{A}$, we write $A \in_{\mathbb{Q}\mathcal{A}} M \oplus K$ for the biproduct of M and K in $\mathbb{Q}\mathcal{A}$.

Proposition 3.13. *Let A, M, and K be torsion-free abelian groups.*

1) *Let $nA \subseteq M \oplus K \subseteq A$ in $\mathcal{A}$ for some positive integer n. Then $A \in_{\mathbb{Q}\mathcal{A}} M \oplus K$. In particular, if $A = M \oplus K$ in $\mathcal{A}$, then $A \in_{\mathbb{Q}\mathcal{A}} M \oplus K$.*

2) *Suppose that $A \in_{\mathbb{Q}\mathcal{A}} M \oplus K$. Then there are a positive integer n and subgroups M', K' of A such that $M' \cong M$, $K' \cong K$ and $nA \subseteq M' \oplus K' \subseteq A$.*

Proof. 1) By hypothesis and Proposition 3.7 we have $A \cong_{\mathrm{qu}} M \oplus K$, and hence $A \in_{\mathbb{Q}\mathcal{A}} M \oplus K$.

2) Let $\iota_M \in \mathbb{Q}\operatorname{Hom}(M, A)$, $\pi_M \in \mathbb{Q}\operatorname{Hom}(A, M)$, $\iota_{\bar{M}} \in \mathbb{Q}\operatorname{Hom}(M, A)$, and $\pi_M \in \mathbb{Q}\operatorname{Hom}(A, M)$ be a set of structural maps belonging to the biproduct $A \in_{\mathbb{Q}\mathcal{A}} M \oplus K$. Then there is a positive integer m such that $m\iota_M \in \operatorname{Hom}(M, A)$, $m\pi_M \in \operatorname{Hom}(A, M)$, $m\iota_M \in \operatorname{Hom}(M, A)$, and $m\pi_M \in \operatorname{Hom}(A, M)$. Let $n = m^2$, $M' = (m\iota_M)(M)$ and $K' = (m\iota_K)(K)$. Then it follows from $\iota_M\pi_M + \iota_K\pi_K = 1_{\mathbb{Q}\mathcal{A}}$ that $(m\iota_M)(m\pi_M) + (m\iota_K)(m\pi_K) = m^2 \cdot 1_A$, and therefore for every $x \in A$,

$$m^2 x = (m\iota_M)(m\pi_M)(x) + (m\iota_K)(m\pi_K)(x) \in M' \oplus K'.$$

Hence $m^2 A \subseteq M' + K' \subseteq A$. Suppose that $x \in M' \cap K'$. Then $x = (m\iota_M)(h) = (m\iota_K)(k)$ for some $h \in M$ and some $k \in K$. Using that $\pi_K\iota_M = 0$ and $\pi_M\iota_K = 0$ we obtain

$$m^2 x = (m\iota_M)(m\pi_M)((m\iota_K)(k)) + (m\iota_K)(m\pi_K)((m\iota_M)(h)) = 0.$$

Hence $M' \cap K' = 0$. Finally, the maps $M \ni x \mapsto (m\iota_M)(x) \in M'$ and $K \ni x \mapsto (m\iota_K)(x) \in K'$ are isomorphisms because they are surjective by definition and also injective because, e.g., if $h \in M$ and $m\iota_M(h) = 0$, then $0 = (m\pi_M)(m\iota_M)(h) = m^2 \cdot 1_M(h) = m^2 h$, so $h = 0$. $\qquad\square$

Definition 3.14. We call $A \in_{\mathbb{Q}\mathcal{A}} M \oplus K$ a **quasi-decomposition** of A. A group M is called a **quasi-summand** of A if and only if there exists a quasi-decomposition $A \in_{\mathbb{Q}\mathcal{A}} M \oplus K$. An indecomposable group of $\mathbb{Q}\mathcal{A}$ is called **strongly indecomposable**.

Strongly indecomposable groups are plentiful. Simple examples are the groups of rank 1.

Example 3.15. Let A be an additive subgroup of $\mathbb{Q}$, so $\operatorname{rk}(A) = 1$. Then A is strongly indecomposable because of rank. It is well-known that $\operatorname{End}(A) = \{\frac{m}{n} : nA = A\}$ where the fraction $\frac{m}{n}$ is assumed to be reduced and $\frac{m}{n}$ acts by multiplication on elements of A. Hence $\mathbb{Q}\operatorname{End}(A) = \mathbb{Q}$.

We will now give an example showing that indecomposable groups need not be strongly indecomposable.

Example 3.16. The group $X = \left(\mathbb{Z}[2^{-1}]v_1 \oplus \mathbb{Z}[3^{-1}]v_2\right) + \mathbb{Z}\frac{1}{5}(v_1 + v_2)$ is indecomposable but not strongly indecomposable as $X \in_{\mathbb{Q}\mathcal{A}} \mathbb{Z}[2^{-1}]v_1 \oplus \mathbb{Z}[3^{-1}]v_2$.

Proof. Clearly $5X \subseteq \mathbb{Z}[2^{-1}]v_1 \oplus \mathbb{Z}[3^{-1}]v_2 \subseteq X$, hence $X \in_{\mathbb{Q}\mathcal{A}} \mathbb{Z}[2^{-1}]v_1 \oplus \mathbb{Z}[3^{-1}]v_2$ by Proposition 3.13. We show next that X is indecomposable. Note that every element $x \in X$ is of the form

$$x = \left(\frac{a_1}{2^{a_2}} + \frac{m}{5}\right)v_1 + \left(\frac{b_1}{3^{b_2}} + \frac{m}{5}\right)v_2$$

where $m \in \mathbb{Z}$, a_1 is an odd integer, a_2 is a non-negative integer, b_1 is an integer not divisible by 3, and b_2 is a non-negative integer. Using this representation it is not hard to show that $\mathbb{Z}[2^{-1}]v_1$ is the unique maximal 2-divisible subgroup of X and $\mathbb{Z}[3^{-1}]v_2$ is the unique maximal 3-divisible subgroup of X. Both these groups are pure fully invariant subgroups of X and indecomposable (having rank 1).

Now suppose that $X = Y \oplus Z$ for subgroups Y and Z of X. By Lemma 2.5 we conclude that $\mathbb{Z}[2^{-1}]v_1 = Y \cap \mathbb{Z}[2^{-1}]v_1 \oplus Z \cap \mathbb{Z}[2^{-1}]v_1$. As $\mathbb{Z}[2^{-1}]v_1$ is indecomposable we assume without loss of generality that $Y \cap \mathbb{Z}[2^{-1}]v_1 = \mathbb{Z}[2^{-1}]v_1$, i.e., $\mathbb{Z}[2^{-1}]v_1 \subseteq Y$. Similarly, either $\mathbb{Z}[3^{-1}] \subseteq Y$ or $\mathbb{Z}[3^{-1}] \subseteq Z$. If Y contains both $\mathbb{Z}[2^{-1}]v_1$ and $\mathbb{Z}[3^{-1}]v_2$, then $5X \subseteq Y$ and $Y = Y_*^X = X$. In this case we are done. The other possibility is that $\mathbb{Z}[2^{-1}]v_1 \subseteq Y$ and $\mathbb{Z}[3^{-1}]v_2 \subseteq Z$. In this case $5\mathbb{Z}[2^{-1}]v_1 \oplus 5\mathbb{Z}[3^{-1}]v_2 = 5X \subseteq \mathbb{Z}[2^{-1}]v_1 \oplus \mathbb{Z}[3^{-1}]v_2 \subseteq X = Y \oplus Z$ and it follows that $5Y \subseteq \mathbb{Z}[2^{-1}]v_1 \subseteq Y$ and $5Z \subseteq \mathbb{Z}[3^{-1}]v_2 \subseteq Z$ and so $Y = \mathbb{Z}[2^{-1}]v_1$ and $Z = \mathbb{Z}[3^{-1}]v_2$. This is a contradiction because then $\frac{1}{2}(v_1 + v_2) \notin Y \oplus Z = X$. $\quad\square$

Remark. The groups in Example 3.1 and Example 3.16 are examples of almost completely decomposable groups. A group is **completely decomposable** if it is a direct sum of rank-1 groups. A torsion-free group of finite rank is **almost completely decomposable** if it contains a completely decomposable subgroup of finite index. As the examples show such groups are fairly concrete and can be approached computationally, but the computations become very nasty very soon. Luckily there is by now an extensive theory of almost completely decomposable groups that goes a long way although it also shows that these groups are very complicated. See the survey article [22] or the comprehensive book [23].

Lemma 3.17. *Idempotents split in* $\mathbb{Q}\mathcal{A}$. *Moreover, if* $A \in \mathbb{Q}\mathcal{A}$, $E^2 = E \in \mathbb{Q}\operatorname{End}(A)$ *and* $e = nE \in \operatorname{End}(A)$, *then* $nA \subseteq \operatorname{Ker}(e) \oplus \operatorname{Im}(e) \subseteq A$, *i.e.,* $A \in_{\mathbb{Q}\mathcal{A}} \operatorname{Ker}(e) \oplus \operatorname{Im}(e)$.

Proof. Let $A \in \mathbb{Q}\mathcal{A}$, and $E^2 = E \in \mathbb{Q}\operatorname{End}(A)$. Then we take E to be an idempotent linear transformation $E : \mathbb{Q}A \to \mathbb{Q}A$ and we get $\mathbb{Q}A = E(\mathbb{Q}A) \oplus (1 - E)(\mathbb{Q}A)$ with structural maps $\iota : E(\mathbb{Q}A) \ni x \mapsto x \in \mathbb{Q}A$ and $\pi : \mathbb{Q}A \ni x \mapsto E(x) \in E(\mathbb{Q}A)$. We have

$$\pi\iota = 1_{E(\mathbb{Q}A)} \quad \text{and} \quad \iota\pi = E. \tag{7}$$

As E is a quasi-endomorphism of A, there exists a positive integer n such that $e := nE \in \operatorname{End}(A)$. Then $(\iota \restriction_{eA}) : eA \to A$ is a well-defined homomorphism because $\iota(x) = x$, and $n(\pi \restriction_A) : A \to eA$ is a well-defined homomorphism because $n\pi(A) = nE(A) = e(A) \subseteq A$, so $(\pi \restriction_A) \in \mathbb{Q}\operatorname{Hom}(A, eA)$. By (7) we have that $(\pi \restriction_A)(\iota \restriction_{eA}) = 1_{eA}$ and $(\iota \restriction_{eA})(\pi \restriction_A) = E$ and this shows that E splits in $\mathbb{Q}\mathcal{A}$.

Observe that $ne = e^2$ and so $(n - e)e = 0$. We claim that

$$nA \subseteq \operatorname{Ker}(e) \oplus e(A) \subseteq A.$$

In fact, let $x \in A$ be given. Then $nx = (n - e)(x) + e(x)$, hence $nA \subseteq (n - e)(A) + e(A) \subseteq A$. Also $(n - e)(A) \subseteq \operatorname{Ker}(e)$ because $e(n - e) = 0$. So $nA \subseteq \operatorname{Ker}(e) + \operatorname{Im}(e) \subseteq A$. The sum is direct because $eA = nEA \subseteq E(\mathbb{Q}A)$ and $\operatorname{Ker}(e) = \operatorname{Ker}(nE) \subseteq \operatorname{Ker}(E) = (1 - E)\mathbb{Q}A$. $\quad\square$

The ring $\mathbb{Q}\operatorname{End}(A)$ for $A \in \mathbb{Q}\mathcal{A}$ is a finite-dimensional $\mathbb{Q}$-algebra, hence has finite length.

Proposition 3.18. *Let* $A \in \mathbb{Q}\mathcal{A}$. *Then* A *is indecomposable in* $\mathbb{Q}\mathcal{A}$ *if and only if* $\mathbb{Q}\operatorname{End}(A)$ *is local.*

Proof. (Arnold [3]) Let A be an indecomposable object in $\mathcal{Q}\mathcal{A}$. Then $\mathbb{Q}\operatorname{End}(A)$ contains no idempotents other than 0 and 1 since idempotents split in $\mathcal{Q}\mathcal{A}$. Also $\mathbb{Q}\operatorname{End}(A)$ is Artinian and therefore its radical $\operatorname{Rad}(\mathbb{Q}\operatorname{End}(A))$ is nilpotent. It follows that $\mathbb{Q}\operatorname{End}(A)/\operatorname{Rad}(\mathbb{Q}\operatorname{End}(A))$ contains no idempotents other than 0 and 1, since idempotents lift modulo the nilpotent radical $\operatorname{Rad}(\mathbb{Q}\operatorname{End}(A))$. Thus the quotient ring $\mathbb{Q}\operatorname{End}(A)/\operatorname{Rad}(\mathbb{Q}\operatorname{End}(A))$ is a semisimple Artinian ring without proper idempotents and therefore a division ring. Consequently, $\mathbb{Q}\operatorname{End}(A)$ is local. Conversely, a local ring contains no proper idempotents, hence A is indecomposable. $\qquad\square$

The existence of indecomposable decompositions in $\mathcal{Q}\mathcal{A}$ is guaranteed by rank arguments. The uniqueness of decompositions is a consequence of the fact that strongly indecomposable groups have local quasi-endomorphism rings. The proper tool is provided by the so-called "Azumaya Unique Decomposition Theorem". A detailed proof is given in [23] for a preadditive category because this version was needed in the context of almost completely decomposable groups. The Azumaya Theorem is usually proved for additive categories. The difference between additive and preadditive categories is the lack of a biproduct for every given finite set of objects in the latter, but it is intuitively clear that the uniqueness question that deals with two given biproducts should not need the existence of arbitrary biproducts, only the existence of subsums.

Theorem 3.19 (Azumaya Unique Decomposition Theorem). *Let $\mathcal{C}$ be a preadditive category in which idempotents split. Suppose that $A \in A_1 \oplus \cdots \oplus A_m$ and also $A \in B_1 \oplus \cdots \oplus B_n$.*

1) *If each $\operatorname{Hom}_{\mathcal{C}}(A_i, A_i)$ is a local ring, then, for, for every j there exist objects $B_{j1}, \ldots, B_{js}$ such that $B_j \in B_{j1} \oplus \cdots \oplus B_{js}$ and each B_{jk} is isomorphic with one of the A_i.*

2) *If, in addition, each of the B_j is indecomposable, then $n = m$ and, after relabeling if necessary, $B_i \cong A_i$ for $i \in \{1, \ldots, n\}$.*

Corollary 3.20. *If $A \in_{\mathcal{Q}\mathcal{A}} A_1 \oplus \cdots \oplus A_m$ and also $A \in_{\mathcal{Q}\mathcal{A}} B_1 \oplus \cdots \oplus B_n$ are indecomposable decompositions in $\mathcal{Q}\mathcal{A}$, then $m = n$ and, after relabeling if necessary, $A_i \cong_{\mathrm{qu}} B_i$ for every i.*

Before getting into regularity we look at kernels and images in the category $\mathcal{Q}\mathcal{A}$. We are confronted with a notational conflict. For a morphism $F : A \to B$ in $\mathcal{Q}\mathcal{A}$ we have the category theoretical $\operatorname{Ker}(F)$ and $\operatorname{Im}(F)$ and we also have the usual kernel and image of the linear transformation $F : \mathbb{Q}A \to \mathbb{Q}B$. We will denote the latter by $\operatorname{Ker}_{\mathbb{Q}}(F)$ and $\operatorname{Im}_{\mathbb{Q}}(F)$. For a map $f \in \operatorname{Hom}(A, M) \subseteq \mathbb{Q}\operatorname{Hom}(A, M)$, the symbols $\operatorname{Ker}_{\mathcal{A}}(f)$ and $\operatorname{Im}_{\mathcal{A}}(f)$ will be the usual kernel and image.

The connections between the usual kernels and images are as follows.

Lemma 3.21. *Let $F \in \mathbb{Q}\operatorname{Hom}(A, M)$ and let $0 \neq n \in \mathbb{N}$ be such that $f := nF \in \operatorname{Hom}(A, M)$. Let*

$$\operatorname{Ker}_{\mathbb{Q}}(F) := \{v \in \mathbb{Q}A \mid F(v) = 0\}, \quad and \quad \operatorname{Im}_{\mathbb{Q}}(F) := F(\mathbb{Q}A),$$

i.e., $\operatorname{Ker}_\mathbb{Q}(F)$ *and* $\operatorname{Im}_\mathbb{Q}(F)$ *are the kernel and image of the linear transformation* $F : \mathbb{Q}A \to \mathbb{Q}M$. *Let*

$$\operatorname{Ker}_\mathcal{A}(f) := \{x \in A \mid f(x) = 0\}, \quad \text{and} \quad \operatorname{Im}_\mathcal{A}(f) := f(A),$$

i.e., $\operatorname{Ker}_\mathcal{A}(f)$ *and* $\operatorname{Im}_\mathcal{A}(f)$ *are the kernel and image of the homomorphism* $f : A \to M$. *Then*

$$\operatorname{Ker}_\mathcal{A}(f) = A \cap \operatorname{Ker}_\mathbb{Q}(F), \quad \text{and} \quad \mathbb{Q}\operatorname{Ker}_\mathcal{A}(f) = \operatorname{Ker}_\mathbb{Q}(F),$$

$$\mathbb{Q}\operatorname{Im}_\mathcal{A}(f) = \operatorname{Im}_\mathbb{Q}(F), \quad \text{and} \quad \operatorname{Im}_\mathcal{A}(f) \subseteq A \cap \operatorname{Im}_\mathbb{Q}(F) = (\operatorname{Im}(f))_*^M.$$

Proof. Let $x \in \operatorname{Ker}_\mathcal{A}(f)$. Then $nF(x) = f(x) = 0$, hence $F(x) = 0$. Thus $\operatorname{Ker}_\mathcal{A}(f) \subseteq A \cap \operatorname{Ker}_\mathbb{Q}(F)$. Let $x \in A \cap \operatorname{Ker}_\mathbb{Q}(F)$. Then $0 = nF(x) = f(x)$ so $x \in \operatorname{Ker}_\mathcal{A}(f)$. Next, let $x \in \mathbb{Q}A$ and assume that $F(x) = 0$. There is $0 \neq m \in \mathbb{N}$ such that $mx \in A$. Then $mx \in A \cap \operatorname{Ker}_\mathbb{Q}(F) = \operatorname{Ker}_\mathcal{A}(f)$. This shows that $\operatorname{Ker}_\mathbb{Q}(F) \subseteq \mathbb{Q}\operatorname{Ker}_\mathcal{A}(f)$. The containment $\mathbb{Q}\operatorname{Ker}_\mathcal{A}(f) \subseteq \operatorname{Ker}_\mathbb{Q}(F)$ follows from the fact that $\operatorname{Ker}_\mathcal{A}(f) \subseteq \operatorname{Ker}_\mathbb{Q}(F)$ and $\operatorname{Ker}_\mathbb{Q}(F)$ is a $\mathbb{Q}$-vector space.

Turning to images, let $x \in \operatorname{Im}_\mathbb{Q}(F)$. Then there is $y \in \mathbb{Q}A$ such that $x = F(y)$. Choose $0 \neq m \in \mathbb{N}$ such that $my \in A$. Then $nmx = nmF(y) = (nF)(my) = f(my) \in \operatorname{Im}_\mathcal{A}(f)$. Hence $\mathbb{Q}f(A) \subseteq F(\mathbb{Q}A) \subseteq \mathbb{Q}f(A)$, and consequently, $f(A) \subseteq A \cap F(\mathbb{Q}A) = A \cap \mathbb{Q}f(A) = (f(A))_*^A$. $\qquad\square$

We will use Definition 2.1.

Lemma 3.22. *Morphisms in* $\mathbb{Q}A$ *have kernels and cokernels. More precisely, let* $F \in \mathbb{Q}\operatorname{Hom}(A, B)$ *and let* $f := mF \in \operatorname{Hom}(A, B)$ *for* $0 \neq m \in \mathbb{N}$.

1) *Let* $\iota : \operatorname{Ker}_\mathcal{A}(f) \to A$ *be the insertion in* $\mathcal{A}$. *Then* $\iota \in \operatorname{Ker}(F)$.

2) *Let* $\nu : B \to B/(\operatorname{Im}(f))_*^B$ *be the natural epimorphism in* $\mathcal{A}$. *Then* $\nu \in \operatorname{Coker}(F)$.

Proof. We will use Lemma 3.6 without explicit reference.

1) Note that ι is monic. From $f\iota = 0$, it follows that $F\iota = 0$. Suppose that $\Xi \in \mathbb{Q}\operatorname{Hom}(X, A)$ such that $F\Xi = 0$. Let $0 \neq n \in \mathbb{N}$ be such that $\xi := n\Xi \in \operatorname{Hom}(X, A)$. Then $f\xi = (mF)(n\Xi) = mnF\Xi = 0$, and hence $\operatorname{Im}_\mathcal{A}\xi \subseteq \operatorname{Ker}_\mathcal{A}(f)$ and $\tilde{\xi} : X \ni x \mapsto \xi(x) \in \operatorname{Ker}_\mathcal{A}(f)$ is in $\operatorname{Hom}(X, \operatorname{Ker}_\mathcal{A}(f))$. Let $\tilde{\Xi} = \frac{1}{n}\tilde{\xi}$. Then $\iota\tilde{\xi} = \xi$, so also $\iota\tilde{\Xi} = \Xi$. Uniqueness: Suppose that $\iota\Xi_1 = G$ and $\iota\Xi_2 = G$. Then $\iota\Xi_1 = \iota\Xi_2$ and ι being monic it follows that $\Xi_1 = \Xi_2$.

2) Note that ν is epic. From $\nu f = 0$, it follows that $\nu F = 0$. Let $\Xi \in \mathbb{Q}\operatorname{Hom}(B, X)$ be such that $\Xi F = 0$. Then $\Xi(F(\mathbb{Q}A)) = 0$. Define

$$\Phi : B/(\operatorname{Im}(f))_*^B \to X : \Phi(\nu(b)) = \Xi(b), \text{ where } b \in B.$$

We will show that Φ is a well-defined map in $\mathbb{Q}\operatorname{Hom}(B/(\operatorname{Im} f)_*^B, X)$ and then by definition, $\Phi\nu = \Xi$.

Suppose that $\nu(b) = \nu(b')$. Then $b - b' \in (\operatorname{Im}_{\mathcal{A}} f)^B_* = A \cap F(\mathbb{Q}A)$ and as $\Xi(F(\mathbb{Q}A)) = 0$ it follows that $\Xi(b) - \Xi(b') = 0$. Hence Φ is well-defined as a linear transformation.

To show that Φ is a quasi-homomorphism, choose $0 \neq n \in \mathbb{N}$ such that $n\Xi \in \operatorname{Hom}(B, X)$. Then $(n\Phi)(\nu(B)) = n\Xi(B) \subseteq X$. Uniqueness follows from the fact that ν is epic. $\qquad\square$

A category is **preabelian** if it is additive and has kernels and cokernels.

Proposition 3.23. *The quasi-isomorphism category $\mathbb{Q}A$ is a preabelian category.*

We will give an example showing that $\mathbb{Q}A$ is not an abelian category.
To do so we need a lemma from category theory.

Lemma 3.24. *Let $f : A \to M$ be a morphism, assume that kernels and cokernels exist as needed and let*

$$(k : K \to A) \in \operatorname{Ker}(f), \quad (c : M \to C) \in \operatorname{Coker}(f),$$

$$(c' : A \to C') \in \operatorname{Coker}(k) = \operatorname{Coim}(f), (k' : K' \to M) \in \operatorname{Ker}(c) = \operatorname{Im}(f).$$

Then there is a unique morphism $\overline{f} : C' \to K'$ such that $f = k'\overline{f}c'$.

$$
\begin{array}{ccccccc}
K & \xrightarrow{k} & A & \xrightarrow{f} & M & \xrightarrow{c} & C \\
 & & c' \downarrow & & \uparrow k' & & \\
 & & C' & \xrightarrow{\overline{f}} & K' & &
\end{array}
$$

Proof. The uniqueness follows because k' is monic and c' is epic and therefore these maps can be canceled in $k'\overline{f}c' = k'\overline{\overline{f}}c'$. For the existence we use that $cf = 0$ and $k' \in \operatorname{Ker}(c)$ to get a morphism $f_1 : A \to K'$ such that $f = k'f_1$. We now find that $k'f_1 k = fk = 0$ and since k' is monic, already $f_1 k = 0$. But $c' \in \operatorname{Coker}(k)$ hence there is $f_2 : C' \to K'$ such that $f_1 = f_2 c'$. The morphism $\overline{f} := f_2$ is the desired map. $\qquad\square$

For a preabelian category to be abelian it is required that for every morphism $f \in \mathcal{C}(A, M)$ the induced map $\overline{f} : C' \to K'$ is an isomorphism. Let us clear up what the induced map is in our case.

By definition $\operatorname{Im}(F) = \operatorname{Ker}(\operatorname{Coker}(F))$ and $\operatorname{Coim}(F) = \operatorname{Coker}(\operatorname{Ker}(F))$. Let $f := mF \in \operatorname{Hom}(A, B)$. Then (Lemma 3.22)

$$(\iota : \operatorname{Ker}_{\mathcal{A}}(f) \to A) \in \operatorname{Ker}(F), \quad \text{and} \quad (\nu : B \to B/(\operatorname{Im}_{\mathcal{A}} f)^B_*) \in \operatorname{Coker}(F).$$

We now see that for the insertion ins and the natural epimorphism nat

$$(\text{ins} : (\operatorname{Im}_{\mathcal{A}} f)^B_* \to B) \in \operatorname{Im}(F) \quad \text{and} \quad (\text{nat} : A \to A/\operatorname{Ker}_{\mathcal{A}}(f)) \in \operatorname{Coim}(F)$$

and we have the induced map

$$\overline{F} : A/\operatorname{Ker}_{\mathcal{A}}(f) \ni a + \operatorname{Ker}_{\mathcal{A}}(f) \mapsto f(a) \in (\operatorname{Im}_{\mathcal{A}} f)^B_*.$$

As $A/\operatorname{Ker}_{\mathcal{A}}(f) \cong \operatorname{Im}_{\mathcal{A}} f$ it is evident that $\overline{F}$ need not be surjective although it is injective and there may not be an inverse of $\overline{F}$ in $\mathbb{Q}\mathcal{A}$. We give a concrete example.

Example 3.25. Let $F : \mathbb{Z} \oplus \mathbb{Z} \ni (a,b) \mapsto a - b \in \mathbb{Q}$. Then $\overline{F} : 0 \to \mathbb{Q}$ which is not invertible.

Proof. In our example the diagram in Lemma 3.24 becomes

$$
\begin{array}{ccccccc}
\operatorname{Ker}_{\mathcal{A}}(f) & \xrightarrow{\text{ins}} & \mathbb{Z} \oplus \mathbb{Z} & \xrightarrow{F} & \mathbb{Q} & \xrightarrow{0} & 0 \\
& & 0 \downarrow & & \uparrow 1_{\mathbb{Q}} & & \\
& & 0 & \xrightarrow{\overline{F}} & \mathbb{Q} & &
\end{array}
$$

and $\overline{F} : 0 \to \mathbb{Q}$ does not have an inverse. $\qquad\square$

We now steer toward exact sequences in $\mathbb{Q}\mathcal{A}$. We settle first the question which quasi-homomorphisms are monic and which are epic.

Lemma 3.26. *Let $F \in \mathbb{Q}\operatorname{Hom}(A,B)$, and $n \neq 0$ such that $f = nF \in \operatorname{Hom}(A,B)$.*

1) *F is monic if and only if f is injective if and only if $F \in \operatorname{Hom}(\mathbb{Q}A, \mathbb{Q}B)$ is injective as a linear transformation.*

2) *F is epic if and only if $B/f(A)$ is torsion if and only if $F(\mathbb{Q}A) = \mathbb{Q}B$, i.e., if and only if F is surjective as a linear transformation.*

Proof. 1) Suppose that $U, V \in \mathbb{Q}\operatorname{Hom}(C,A)$ and $FU = FV$. There is $n \neq 0$ such that $f := nF \in \operatorname{Hom}(A,B)$, $u := nU, v := nV \in \operatorname{Hom}(C,A)$. Then $fu = fv$. If f is injective, then $u = v$ and hence also $U = V$, so F is monic.

Now suppose that F is monic and choose $n \neq 0$ such that $f := nF \in \operatorname{Hom}(A,B)$. Assume by way of contradiction that f is not injective. Let $u : \operatorname{Ker}_{\mathcal{A}}(f) \to A$ be the insertion and let $v : \operatorname{Ker}_{\mathcal{A}}(f) \to A$ be the zero map. Then $fu = 0 = fv$ and also $Fu = Fv$. By cancellation we obtain the contradiction $u = v$.

Finally, if $f = nF$ for some nonzero integer n, then nF considered as the unique linear transformation that extends f to $\mathbb{Q}A$ is injective if and only if f is injective, and nF is injective if and only if F is injective.

2) Suppose that F is epic. Then $B/(f(A))_*^B$ is either 0 or $B/(f(A))_*^B$ is a nonzero torsion-free group. Assume by way of contradiction that the latter is the case. Then we have map $u := 0 : B \to B/(f(A))_*^B$ and $0 \neq v : B \ni b \mapsto b + (f(A))_*^B \in B/(f(A))_*^B$. It follows that $uF = vF$ but F cannot be canceled, a contradiction.

Now suppose that F is surjective as a linear transformation. Then it can be canceled on the right. $\qquad\square$

We can now deal with exact sequences. We first recall the definition.

Definition 3.27. Let $E : A \xrightarrow{f} B \xrightarrow{g} C$ be a sequence of morphisms in a category with kernels and cokernels. The sequence is **exact at** B if $\mathrm{Im}(f) = \mathrm{Ker}(g)$. i.e., if $(c : B \to C) \in \mathrm{Coker}(f)$ and $(k : K \to B) \in \mathrm{Ker}(c)$, so $k \in \mathrm{Im}(f)$, then E is exact at B if and only if $k \in \mathrm{Ker}(g)$.

$$
\begin{array}{ccccc}
 & & K & & \\
 & & k \downarrow & & \\
A & \xrightarrow{f} & B & \xrightarrow{g} & C \\
 & & c \downarrow & & \\
 & & C & &
\end{array}
$$

Proposition 3.28. *Let* $E : A \xrightarrow{F} B \xrightarrow{G} C$ *be a sequence of morphisms in* $Q\mathcal{A}$. *Choose nonzero integers* m, n *such that* $f := mF \in \mathcal{A}$ *and* $g := nG \in \mathcal{A}$. *Then the following statements are equivalent.*

1) *E is exact at B.*

2) *$F(A) \cap B \subseteq \mathrm{Ker}_{\mathbb{Q}}(G)$ and $(\mathrm{Ker}_{\mathbb{Q}}(G) \cap B)/(F(A) \cap B)$ is bounded (so finite).*

3) *There exist integers e_1 and e_2 such that $e_1 \mathrm{Im}(f) \subseteq \mathrm{Ker}(g)$ and $e_2 \mathrm{Ker}(g) \subseteq \mathrm{Im}(f)$, i.e., $\mathrm{Im}(f)$ and $\mathrm{Ker}(g)$ are "quasi-equal".*

Proof. 1) $\Leftrightarrow$ 2). By Lemma 3.22 we know kernels and cokernels in the quasi-isomorphism category. The relevant diagram is as follows.

$$
\begin{array}{ccccc}
 & & F(A) \cap B & & \\
 & & \mathrm{ins} \downarrow & & \\
A & \xrightarrow{F} & B & \xrightarrow{G} & C \\
 & & \mathrm{nat} \downarrow & & \\
 & & B/(F(A) \cap B) & &
\end{array}
$$

By definition E is exact at B if and only if $\mathrm{ins} \in \mathrm{Ker}(G)$.

Suppose that E is exact. Then $G \circ \mathrm{ins} = 0$ which is equivalent to $F(A) \cap B \subseteq \mathrm{Ker}_{\mathbb{Q}}(G)$. We also have the insertion $\iota : \mathrm{Ker}_{\mathbb{Q}}(G) \cap B \to B$ with the property that $G\iota = 0$. Using that $\mathrm{ins} \in \mathrm{Ker}(G)$ we obtain a morphism $\Phi : \mathrm{Ker}_{\mathbb{Q}}(G) \cap B \to F(A) \cap B$ such that $\iota = \mathrm{ins} \circ \Phi$. There exists $0 \neq m \in \mathbb{N}$ such that $\phi := m\Phi \in \mathrm{Hom}(\mathrm{Ker}_{\mathbb{Q}}(G) \cap B, F(A) \cap B)$. Then $m(\mathrm{Ker}_{\mathbb{Q}}(G) \cap B) = m\iota(\mathrm{Ker}_{\mathbb{Q}}(G) \cap B) = m \, \mathrm{ins} \, \Phi(\mathrm{Ker}_{\mathbb{Q}}(G) \cap B) = \phi((\mathrm{Ker}_{\mathbb{Q}}(G) \cap B) \subseteq F(A) \cap B$.

Conversely, assume that $F(A) \cap B \subseteq \mathrm{Ker}_{\mathbb{Q}}(G)$ and for some positive integer $m(\mathrm{Ker}_{\mathbb{Q}}(G) \cap B) \subseteq F(A) \cap B$. We must show that $\mathrm{ins} \in \mathrm{Ker}(G)$. The first condition implies at once that $G \circ \mathrm{ins} = 0$. Now suppose that $\Phi : X \to B$ is a morphism such that $G\Phi = 0$. We must show that Φ factors through ins. Let n be a positive integer such that $\phi = n\Phi \in \mathrm{Hom}(X, B)$. Then $G\phi = 0$ hence $\phi(X) \subseteq \mathrm{Ker}_{\mathbb{Q}}(G) \cap B$ and $nm\Phi(X) = m\phi(X) \subseteq m(\mathrm{Ker}_{\mathbb{Q}}(G) \cap B) \subseteq F(A) \cap B$. This means that $\Psi : \mathbb{Q}X \ni x \mapsto \Phi(x) \in \mathbb{Q}(F(A) \cap B)$ is a morphism $\Psi \in \mathbb{Q}\mathrm{Hom}(X, F(A) \cap B)$. Clearly, $\Phi = \mathrm{ins} \circ \Psi$ which was needed.

2) $\Leftrightarrow$ 3). $F(A) \cap B \subseteq \mathrm{Ker}_{\mathbb{Q}}(G)$ and $\mathrm{Ker}_{\mathbb{Q}}(G) \cap B / F(A) \cap B$ is bounded is equivalent with

$$F(A) \cap B \subseteq \mathrm{Ker}_{\mathcal{A}}(g) \text{ and } e\,\mathrm{Ker}_{\mathbb{Q}}(G) \cap B \subseteq F(A) \text{ for some } 0 \neq e \in \mathbb{N},$$

which in turn is equivalent with

$$f(A) = mF(A) \cap B \subseteq F(A) \cap B \subseteq \mathrm{Ker}_{\mathcal{A}}(g) \text{ and } me\,\mathrm{Ker}_{\mathcal{A}}(g) \subseteq m(F(A) \cap B) \subseteq$$
$$mF(A) \cap B = f(A). \qquad \Box$$

We record two special cases.

Corollary 3.29.

1) Let $E : 0 \xrightarrow{F} B \xrightarrow{G} C$ be a sequence of morphisms in $\mathbb{Q}\mathcal{A}$. Then E is exact at B if and only if G is monic, or, equivalently, if and only if $G : \mathbb{Q}B \to \mathbb{Q}C$ is injective as a linear transformation.

2) Let $E : A \xrightarrow{F} B \xrightarrow{G} 0$ be a sequence of morphisms in $\mathbb{Q}\mathcal{A}$. Then E is exact at B if and only if F is epic, or, equivalently, if and only if $B = (\mathrm{Im}_{\mathbb{Q}}(F) \cap B)_*^B$.

Proof. 1) Suppose that E is exact at B. As $F(A) = 0$ in this case it follows that $\mathrm{Ker}_{\mathbb{Q}}(G) \cap B$ is bounded, hence $\mathrm{Ker}_{\mathbb{Q}}(G) \cap B = 0$ and by Lemma 3.21 it follows that $\mathrm{Ker}_{\mathbb{Q}}(G) = 0$ and so G is monic. The converse is immediately seen by looking at Lemma 3.28.

2) Suppose that E is exact at B. As $\mathrm{Ker}_{\mathbb{Q}}(G) = B$ in this case, $B/(F(A) \cap B) = B/(\mathrm{Im}_{\mathbb{Q}}(F) \cap B)$ is bounded. But $\mathrm{Im}_{\mathbb{Q}}(F) \cap B$ is pure in B and therefore $B = \mathrm{Im}_{\mathbb{Q}}(F) \cap B$. Thus $\mathbb{Q}B \subseteq \mathrm{Im}_{\mathbb{Q}}(F)$ which means that $F : \mathbb{Q}A \to \mathbb{Q}B$ is surjective and so epic in $\mathbb{Q}\mathcal{A}$. Again the converse is easily seen. $\qquad \Box$

4 Regularity in $\mathbb{Q}\mathcal{A}$

Let $F \in \mathbb{Q}\,\mathrm{Hom}(A, M)$ be regular and let $G \in \mathbb{Q}\,\mathrm{Hom}(M, A)$ be such that $FGF = F$. Then $E := GF$ is an idempotent in $\mathbb{Q}\,\mathrm{End}(A)$ and $D := FG$ is an idempotent in $\mathbb{Q}\,\mathrm{End}(M)$. Choose $0 \neq n \in \mathbb{N}$ such that $f := nF \in \mathrm{Hom}(A, M)$ and $g := nG \in \mathrm{Hom}(M, A)$. Then $e := n^2 E \in \mathrm{End}(A)$ and $d := n^2 D \in \mathrm{End}(M)$. By Lemma 3.17,

$$n^2 A \subseteq \mathrm{Ker}(e) \oplus \mathrm{Im}(e) \subseteq A, \quad \text{and} \quad n^2 M \subseteq \mathrm{Ker}(d) \oplus \mathrm{Im}(d) \subseteq M.$$

In other words,

$$A \in_{\mathbb{Q}\mathcal{A}} \mathrm{Ker}(e) \oplus \mathrm{Im}(e), \quad \text{and} \quad M \in_{\mathbb{Q}\mathcal{A}} \mathrm{Ker}(d) \oplus \mathrm{Im}(d).$$

Theorem 4.1.

1) Suppose that $0 \neq F \in \mathbb{Q}\,\mathrm{Hom}(A, M)$ is regular and μF is regular for all $\mu \in \mathbb{Q}\,\mathrm{End}(M)$. If M is strongly indecomposable, then $\mathbb{Q}\,\mathrm{End}(M)$ is a division ring.

2) *Suppose that there is* $0 \neq F \in \text{Hom}(A, M)$ *that is regular and* $F\alpha$ *is regular for all* $\alpha \in \mathbb{Q}\text{End}(A)$. *If* A *is strongly indecomposable, then* $\mathbb{Q}\text{End}(A)$ *is a division ring.*

Proof. 1) Let M be strongly indecomposable, assume that $F \in \mathbb{Q}\text{Hom}(A, M)$ is regular, $F \neq 0$, and that μF is regular for every $\mu \in \mathbb{Q}\text{End}(M)$. By hypothesis, we have $FG_1F = F \neq 0$ for some G_1, hence $FG_1 \neq 0$ is a nonzero idempotent of $\mathbb{Q}\text{End}(M)$ and since M is strongly indecomposable and idempotents split in $\mathbb{Q}\mathcal{A}$, it follows that $FG_1 = 1_{\mathbb{Q}M}$. Let $0 \neq \mu \in \mathbb{Q}\text{End}(M)$. Then $\mu FG_1 = \mu \neq 0$, and hence also $\mu F \neq 0$. By hypothesis there is $G \in \mathbb{Q}\text{Hom}(M, A)$ such that $\mu FG\mu F = \mu F \neq 0$. Then μFG is a nonzero idempotent of $\mathbb{Q}\text{End}(M)$ and as M is strongly indecomposable $\mu FG = 1$. Hence every nonzero element in $\mathbb{Q}\text{End}\,M$ has a right inverse and $\mathbb{Q}\text{End}(M)$ is a division ring by Lemma III.1.2.

2) Let A be strongly indecomposable, and assume that $0 \neq F \in \mathbb{Q}\text{Hom}(A, M)$ is regular. Then $FG_1F = F \neq 0$ for some $G_1 \in \mathbb{Q}\text{Hom}(M, A)$. Then G_1F is a nonzero idempotent in $\mathbb{Q}\text{End}(A)$. Since A is strongly indecomposable and idempotents split in $\mathbb{Q}\mathcal{A}$, it follows that $G_1F = 1$. Let $0 \neq \alpha \in \mathbb{Q}\text{End}(A)$. Then $G_1F\alpha = \alpha$ and so $F\alpha \neq 0$. By assumption $F\alpha$ is regular, hence $F\alpha GF\alpha = F\alpha$ for some $G \in \mathbb{Q}\text{Hom}(M, A)$. Hence $GF\alpha$ is a nonzero idempotent of $\mathbb{Q}\text{End}(A)$ and we get $GF\alpha = 1$ and $GF \in \mathbb{Q}\text{End}(A)$ is a left inverse of α. By Lemma III.1.2 the quasi-endomorphism ring $\mathbb{Q}\text{End}(A)$ is a division ring. $\square$

While $\mathbb{Q}$ is up to isomorphism the only group in the category of torsion-free abelian groups whose endomorphism ring is a division ring, there is an abundance of strongly indecomposable groups whose quasi-endomorphism ring is a division ring. There also exist many strongly indecomposable groups whose quasi-endomorphism ring is not a division ring.

There are many results in abelian group theory that state that, given a ring subject to certain conditions, that ring is isomorphic to an endomorphism ring of an abelian group. These results are known as realization theorems. The most celebrated realization theorem is due to A.L.S. Corner.

Theorem 4.2 ([7], [11, Theorem 110.1]). *If* R *is a reduced torsion-free ring with* $\text{rk}(R) = n < \infty$, *then* R *is isomorphic to the endomorphism ring of a torsion-free abelian group of rank* $2n$.

R.S. Pierce and C. Vinsonhaler ([28]) observed the following corollary.

Corollary 4.3. *Let* E *be a finite-dimensional* $\mathbb{Q}$-*algebra with* 1_E. *Then there is a group* A *in* $\mathbb{Q}\mathcal{A}$ *such that* $\mathbb{Q}\text{End}(A) \cong E$.

Proof. In order to apply Corner's Theorem we need to construct a reduced torsion-free ring contained in E. Let $(v_1, \ldots, v_n)$ be a $\mathbb{Q}$-basis of E. Then $v_iv_j = \sum_{k=1}^{n} r_{ijk}v_k$ is a rational linear combination of the basis elements v_k and there exists a positive integer m such that $mr_{ijk} \in \mathbb{Z}$ for all i, j, k. Let $w_i := mv_i$ and let $R := \mathbb{Z}1_E + \mathbb{Z}w_1 + \cdots + \mathbb{Z}w_n$. The R is free as a finitely generated torsion-free abelian group and closed under multiplication because $w_i w_j = m^2 v_i v_j = \sum_{k=1}^{n}(mr_{ijk})mv_k = \sum_{k=1}^{n}(mr_{ijk})w_k \in R$. By Theorem 4.2 there is a group G such that $\mathrm{End}(G) \cong R$ and hence $\mathbb{Q}\,\mathrm{End}(G) \cong \mathbb{Q}R = E$. $\square$

In particular, if F is a finite field extension of $\mathbb{Q}$, then there are groups G in $\mathbb{Q}\mathcal{A}$ with $\mathbb{Q}\,\mathrm{End}(G) \cong F$. Thus the variety of strongly indecomposable groups is huge and little is known about them. On the other hand, $\mathbb{Q}\mathcal{A}$ is a Krull-Schmidt category (Corollary 3.20) and the computation of $\mathrm{Reg}(A, M)$ reduces to determination of $\mathrm{Reg}(A, M)$ for strongly indecomposable groups A and M (Theorem II.6.17) and questions about maps between strongly indecomposable groups. The first question is easy.

Proposition 4.4. *Let A and M be strongly indecomposable groups.*
Then $\mathrm{Reg}(A, M) = 0$ except when $A \cong M$ and $\mathbb{Q}\,\mathrm{End}(A)$ is a division ring in which case $\mathrm{Reg}(A, M) \cong \mathbb{Q}\,\mathrm{End}(A)$.

Proof. Suppose that $f \in \mathbb{Q}\,\mathrm{End}(A, M)$ is regular. Then A and M are quasi-isomorphic by Theorem 2.7 and the strong indecomposability. Finally, $\mathbb{Q}\,\mathrm{End}(A)$ is a division ring by Theorem 4.1. $\square$

An obvious simple case is the following.

Proposition 4.5. *Let $A \in_{\mathbb{Q}\mathcal{A}} A_1 \oplus \cdots \oplus A_m$ and $M \in_{\mathbb{Q}\mathcal{A}} M_1 \oplus \cdots \oplus M_n$ be such that $\mathbb{Q}\,\mathrm{Hom}(A_i, M_j) = 0$ unless $A_i \cong_{\mathrm{qu}} M_j$ and $\mathbb{Q}\,\mathrm{End}(A)$ is a division ring. Then*

$$\mathrm{Reg}(A, M) = \bigoplus \{\mathbb{Q}\,\mathrm{Hom}(A_i, M_j) \mid A_i \cong_{\mathrm{qu}} M_j\}.$$

Unfortunately there may well be non-trivial quasi-homomorphisms between groups whose quasi-endomorphism rings are division rings. Recall that $\mathbb{Q}$ is a strongly indecomposable group whose quasi-endomorphism ring is $\mathbb{Q}$, so even a field.

Example 4.6. Let A be any group in $\mathbb{Q}\mathcal{A}$. Then

$$\mathbb{Q}\,\mathrm{Hom}(A, \mathbb{Q}) \cong \mathbb{Q}^{\mathrm{rk}(A)}.$$

Proof. Here $\mathbb{Q}\,\mathrm{Hom}(A, \mathbb{Q}) = \mathrm{Hom}(A, \mathbb{Q})$. Choose a free subgroup F of A such that A/F is a torsion group and consequently $\mathrm{rk}(F) = \mathrm{rk}(A)$. Then $\mathrm{Hom}(A/F, \mathbb{Q}) = 0$ and since $\mathbb{Q}$ is injective, we have $\mathrm{Hom}(A, \mathbb{Q}) \cong \mathrm{Hom}(F, \mathbb{Q}) \cong \mathbb{Q}^{\mathrm{rk}(F)}$ by standard homological algebra. $\square$

Another easy case is that of completely decomposable groups, i.e., direct sums of rank-1 groups. Note that for any rank-1 group A we have $\mathbb{Q}\,\mathrm{End}(A) = \mathbb{Q}$.

Proposition 4.7. *Let A and M be completely decomposable groups in $\mathbb{Q}\mathcal{A}$. Then $\mathrm{Reg}(A, M) = \mathrm{Hom}(A, M)$.*

In general, little is known about strongly indecomposable groups and homomorphisms between them. Existence theorems for strongly indecomposable groups whose quasi-endomorphism rings are division rings are rarely constructive. We give a construction to illustrate the situation.

4.1 Realizing $\mathbb{Q}(\sqrt{p})$

Let $M_2(\mathbb{Q})$ be the 2×2-matrix ring with rational entries. Fix a prime $p \geq 3$. Note that

$$\begin{bmatrix} 1 & 1 \\ p-1 & -1 \end{bmatrix}^2 = p \begin{bmatrix} 1 & 0 \\ 0 & 1 \end{bmatrix}.$$

Therefore

$$\mathbb{Q} \begin{bmatrix} 1 & 0 \\ 0 & 1 \end{bmatrix} \oplus \mathbb{Q} \begin{bmatrix} 1 & 1 \\ p-1 & -1 \end{bmatrix}$$

is a subalgebra of $M_2(\mathbb{Q})$ that is isomorphic in the obvious way with the field extension $\mathbb{Q}(\sqrt{p})$.

4.2 Constructing the Group

Let $V = \mathbb{Q}v_1 \oplus \mathbb{Q}v_2$ be a $\mathbb{Q}$-vector space, let $\mathbb{P}_p$ be a set of primes and for each $q \in \mathbb{P}_p$ let $c_q \in \mathbb{Z}$, all of these to be specified later. Define

$$G := (\mathbb{Z}v_1 \oplus \mathbb{Z}v_2) + \sum_{q \in \mathbb{P}_p} \mathbb{Z}\tfrac{1}{q}(v_1 + c_q v_2).$$

Set $B := \mathbb{Z}v_1 \oplus \mathbb{Z}v_2$. Then

$$\frac{G}{B} \cong \bigoplus_{q \in \mathbb{P}_p} \frac{\mathbb{Z}\frac{1}{q}(v_1 + c_q v_2) + B}{B} \quad \text{and} \quad \frac{\mathbb{Z}\frac{1}{q}(v_1 + c_q v_2) + B}{B} \cong \frac{\mathbb{Z}}{\mathbb{Z}q}.$$

We now choose $\mathbb{P}_p$ and the c_q. Let

$$\mathbb{P}_p := \{q \in \mathbb{P} \mid q \neq p, q \text{ divides } p - n^2 \text{ for some } n \in \mathbb{N}\}.$$

Lemma 4.8. $\mathbb{P}_p$ *is infinite.*

Proof. Suppose that $q_1, \ldots, q_k \in \mathbb{P}_p$. Then any prime factor of $p - (q_1 q_2 \cdots q_k)^2$ is different from the q_i and from p, hence belongs to $\mathbb{P}_p$. Hence $\mathbb{P}_p$ cannot be a finite set. $\square$

By definition of $\mathbb{P}_p$, for any $q \in \mathbb{P}_p$, the congruence $p \equiv (1+x)^2 \mod q$ has a solution. For each $q \in \mathbb{P}_p$ choose a solution c_q of $p \equiv (1+x)^2 \mod q$. We then have

$$\forall q \in \mathbb{P}_p : p \equiv (1+c_q)^2 \mod q. \tag{8}$$

4.3   Computing the Quasi-Endomorphism Ring

Proposition 4.9.

$$\mathbb{Q}\operatorname{End}(G) = \mathbb{Q}\begin{bmatrix} 1 & 0 \\ 0 & 1 \end{bmatrix} \oplus \mathbb{Q}\begin{bmatrix} 1 & 1 \\ p-1 & -1 \end{bmatrix} \cong \mathbb{Q}(\sqrt{p}).$$

Proof. We first show that G is invariant under the ring

$$\mathbb{Z}\begin{bmatrix} 1 & 0 \\ 0 & 1 \end{bmatrix} \oplus \mathbb{Z}\begin{bmatrix} 1 & 1 \\ p-1 & -1 \end{bmatrix}$$

of linear transformations on V that acts by left multiplication on the coordinate vectors $\begin{bmatrix} x \\ y \end{bmatrix}$ with respect to the basis $\{v_1, v_2\}$. Clearly, we only need to check that the linear transformation $\begin{bmatrix} 1 & 1 \\ p-1 & -1 \end{bmatrix}$ maps each element $\frac{1}{q}(v_1 + c_q v_2) = \frac{1}{q}\begin{bmatrix} 1 \\ c_q \end{bmatrix}$ into G. By a straightforward computation

$$\begin{bmatrix} 1 & 1 \\ p-1 & -1 \end{bmatrix}\frac{1}{q}\begin{bmatrix} 1 \\ c_q \end{bmatrix} = \frac{1+c_q}{q}\begin{bmatrix} 1 \\ c_q \end{bmatrix} + \begin{bmatrix} 0 \\ \frac{1}{q}(p-(1+c_q)^2) \end{bmatrix}$$

where the element $\frac{1}{q}(p-(1+c_q)^2)$ is integral by the choice of q and c_q (see (8)). This shows that

$$\mathbb{Q}\begin{bmatrix} 1 & 0 \\ 0 & 1 \end{bmatrix} \oplus \mathbb{Q}\begin{bmatrix} 1 & 1 \\ p-1 & -1 \end{bmatrix} \subseteq \mathbb{Q}\operatorname{End}(G).$$

To show the converse containment let $\begin{bmatrix} r_{11} & r_{12} \\ r_{21} & r_{22} \end{bmatrix} \in \mathbb{Q}\operatorname{End}(G) \subseteq M_2(\mathbb{Q})$. There exists a positive integer m such that $m\begin{bmatrix} r_{11} & r_{12} \\ r_{21} & r_{22} \end{bmatrix} \in \operatorname{End}(G)$ and all $a_{ij} := mr_{ij}$ are integers. We will show that

$$\alpha := \begin{bmatrix} a_{11} & a_{12} \\ a_{21} & a_{22} \end{bmatrix} = (a_{11} - a_{22})\begin{bmatrix} 1 & 0 \\ 0 & 1 \end{bmatrix} + a_{12}\begin{bmatrix} 1 & 1 \\ p-1 & -1 \end{bmatrix}, \qquad (9)$$

and this shows that $\begin{bmatrix} r_{11} & r_{12} \\ r_{21} & r_{22} \end{bmatrix} \in \mathbb{Q}\operatorname{End}(G)$ as needed. The assumption that $\alpha \in \operatorname{End}(G)$ implies that for every $q \in \mathbb{P}_p$, there exist integers x_q, y_q, z_q such that

$$\begin{bmatrix} a_{11} & a_{12} \\ a_{21} & a_{22} \end{bmatrix}\frac{1}{q}\begin{bmatrix} 1 \\ c_q \end{bmatrix} = \frac{x_q}{q}\begin{bmatrix} 1 \\ c_q \end{bmatrix} + \begin{bmatrix} y_q \\ z_q \end{bmatrix}.$$

This is equivalent to

$$a_{11} + a_{12}c_q = x_q + qy_q, \quad a_{21} + a_{22}c_q = x_q c_q + qz_q,$$

and we get the equivalent system

$$x_q = a_{11} + a_{12}c_q - qy_q, \quad q(z_q - c_qy_q) = a_{21} + c_q(a_{22} - a_{11}) - a_{12}c_q^2.$$

There also is an integer u_q such that $1 + 2c_q + c_q^2 = (1+c_q)^2 = p + qu_q$. Substituting for c_q^2 in the second equation we obtain

$$q(z_q - c_qy_q + a_{12}u_q) = a_{21} - a_{12}(p-1) + c_q(a_{22} - a_{11} + 2a_{12}).$$

Setting $a := a_{21} - a_{12}(p-1)$, $b := a_{22} - a_{11} + 2a_{12}$ and $m_q := z_q - c_qy_q + a_{12}u_q$, then we have the equations

$$qm_q = a + c_qb, \quad p - (1+c_q)^2 = qu_q.$$

Multiplying the second equation by b^2 and substituting c_qb from the first equation we get

$$(a+b)^2 + (p-2)b^2 = q(qm_q^2 - 2am_q + 2bm_q + u_qb^2)$$

which must hold for all $q \in \mathbb{P}_p$. As $\mathbb{P}_p$ is infinite and $p \geq 3$, it follows that $(a+b)^2 + (p-2)b^2 = 0$, $b = a_{22} - a_{11} + 2a_{12} = 0$ and $a = a_{21} - a_{12}(p-1) = 0$. These are the relations needed to validate (9). $\quad\square$

5 Regularity in the Category of Groups

As another example we look at the category of general groups. We begin with the observation that idempotents split in the category of groups in the sense that the existence of an idempotent endomorphism of G implies that G is a semidirect product.

Lemma 5.1. *Let G be a group and let $e = e^2 \in \mathrm{End}(G)$. Then $G = \mathrm{Ker}(e) \rtimes \mathrm{Im}(e)$.*

Proof. Let $x \in G$. Then $x = (xe(x)^{-1})e(x)$ and it is immediate that $xe(x)^{-1} \in \mathrm{Ker}(e)$. Suppose that $e(x) \in \mathrm{Ker}(e)$. Then $1 = e(e(x)) = e(x)$, hence $\mathrm{Ker}(e) \cap \mathrm{Im}(e) = \{1\}$. $\quad\square$

Theorem 5.2. *Let G and H be groups and $f \in \mathrm{Hom}(G,H)$.*

1) *Assume that f is regular. Let $g \in \mathrm{Hom}(H,G)$ such that $fgf = f$. Then $e := gf \in \mathrm{End}(G)$ is an idempotent, $d := fg \in \mathrm{End}(H)$ is an idempotent and*

$$G = \mathrm{Ker}(f) \rtimes \mathrm{Im}(e), \quad \mathrm{Im}(e) \cong \mathrm{Im}(f), \quad and \quad H = \mathrm{Ker}(d) \rtimes \mathrm{Im}(f).$$

2) *Suppose that $G = \mathrm{Ker}(f) \rtimes K$ and $H = N \rtimes \mathrm{Im}(f)$. Then f is regular.*

Proof. 1) That e and d are idempotents follows immediately from the equality $fgf = f$. By Lemma 5.1 we have that $G = \mathrm{Ker}(e) \rtimes \mathrm{Im}(e)$. Also $\mathrm{Ker}(e) = \mathrm{Ker}(f)$ because $e(x) = 1$ if and only if $gf(x) = 1$ if and only if $fgf(x) = f(1) = 1$. We have shown that $G = \mathrm{Ker}(f) \rtimes \mathrm{Im}(e)$ and $\mathrm{Im}(e) \cong G/\mathrm{Ker}(e) \cong \mathrm{Im}(f)$ by the first isomorphism theorem.

Again by Lemma 5.1 we have that $G = \mathrm{Ker}(d) \rtimes \mathrm{Im}(d)$. Clearly $\mathrm{Im}(d) \subseteq \mathrm{Im}(f)$. Suppose that $u \in \mathrm{Ker}(d) \cap \mathrm{Im}(f)$. Then $u = f(x)$ for some $x \in G$, and $u = f(x) = fgf(x) = d(u) = 1$. This shows that $H = \mathrm{Ker}(d) \rtimes \mathrm{Im}(f)$.

2) Suppose that $G = \mathrm{Ker}(f) \rtimes K$ and $H = N \rtimes \mathrm{Im}(f)$. Then $f \restriction_K \colon K \to \mathrm{Im}(f)$ is invertible. Let $\pi : H \to \mathrm{Im}(f)$ be the projection along N which is a homomorphism. Then $g := (f \restriction_K)^{-1}\pi \in \mathrm{Hom}(H, G)$ and $fgf = f$. $\qquad\square$

Bibliography

[1] F. W. Anderson and K. R. Fuller. *Rings and Categories of Modules*, Volume 13 of *Graduate Texts in Mathematics*. Springer Verlag, 1992.

[2] D. M. Arnold. *Finite Rank Torsion Free Abelian Groups and Rings*, volume 931 of *Lecture Notes in Mathematics*. Springer Verlag, 1982.

[3] D. M. Arnold. *Abelian Groups and Representations of Finite Partially Ordered Sets*, volume 2 of *CMS Books in Mathematics*. Springer Verlag, 2000.

[4] B. Brown and N.H. McCoy. The maximal regular ideal of a ring. *Proc. Am. Math. Soc.*, 1:165–171, 1950.

[5] S.U. Chase. Direct products of modules. *Trans. Am. Math. Soc.*, 97:457–473, 1960.

[6] A. L. S. Corner. A note on rank and direct decompositions of torsion-free abelian groups. *Proc. Cambridge Philos. Soc.*, 57:230–33, 1961.

[7] A.L.S. Corner. Every countable reduced torsion-free ring is an endomorphism ring. *Proc. London Math. Soc.*, 13:687–710, 1963.

[8] P. Crawley and R.P. Dilworth. *Algebraic Theory of Lattices*. Prentice Hall, Inc., 1973.

[9] Carl Faith. *Rings and Things and a Fine Array of Twentieth Century Associative Algebra* Mathematical Surveys and Monographs, Volume 65 American Mathematical Society, 1999.

[10] D.J. Fieldhouse. Pure theories. *Math. Ann.*, 184:1–18, 1969.

[11] L. Fuchs. *Infinite Abelian Groups, Vol. I, II.* Academic Press, 1970 and 1973.

[12] L. Fuchs and K.M. Rangaswamy. On generalized regular rings. *Math. Z.*, 107:71–81, 1968.

[13] S. Glaz and W. Wickless. Regular and principal projective endomorphism rings of mixed abelian groups. *Communications in Algebra*, 22:1161–1176, 1994.

[14] K.R. Goodearl. *Von Neumann regular Rings*. Pitman Publishing Limited, 1979.

[15] K.R. Goodearl. Von Neumann regular Rings and Direct Sum Decomposition Problems. In *Abelian Groups and Modules, Proceedings of the 1994 Padova Conference*, Kluwer Academic Publishers, 249–255, 1995.

[16] B. Jonsson. On direct decompositions of torsion-free abelian groups. *Math. Scand.*, 5:230–235, 1957.

[17] B. Jonsson. On direct decompositions of torsion-free abelian groups. *Math. Scand.*, 7:361–371, 1959.

[18] F. Kasch. *Modules and Rings*. A translation from the German "Moduln und Ringe". Academic Press, 1982.

[19] F. Kasch. Regularity in hom. *Algebra Berichte*, 75:1–11, 1996.

[20] F. Kasch and A. Mader. *Rings, Modules, and the Total*. Birkhäuser Verlag, 2004.

[21] F. Kasch and A. Mader. Regularity and substructures of hom. *Communications in Algebra*, 34:1459–1478, 2006.

[22] A. Mader. Almost completely decomposable torsion-free abelian groups. In *Abelian Groups and Modules, Proceedings of the 1994 Padova Conference*, Kluwer Academic Publishers, 343–366, 1995.

[23] A. Mader. *Almost Completely Decomposable Groups*, volume 13 of *Algebra, Logic and Applications*. Gordon and Breach Science Publishers, 2000.

[24] A. Mader. Regularity in endomorphism rings. *Communications in Algebra*, to appear.

[25] S.H. Mohamed and B. Müller. Continuous and discrete modules. *Lecture Notes*, 147, 1990.

[26] W.K. Nicholson. Semiregular modules and rings. *Canadian J. Math.*, 28:1105–1120, 1976.

[27] R.S. Pierce. Modules over Commutative Regular Rings. *Memoirs of the American Math. Soc.*, 70:1–112, 1967.

[28] R.S. Pierce and C. Vinsonhaler. Realizing central division algebras. *Pac. J. Math.*, 109:165–177, 1983.

[29] K.M. Rangaswamy. Abelian groups with special endomorphism rings. *J. Algebra*, 6:271–280, 1967.

[30] K.M. Rangaswamy. Regular and Baer rings. *Proc. Amer. Math. Soc.*, 42:354–358, 1974.

[31] K.M. Rangaswamy and N. Vanaja. A note on modules over regular rings. *Bull. Austral. Math. Soc.*, 4:57–62, 1971.

[32] J. von Neumann. *Continuous Geometry*. Princeton University Press, 1960.

[33] J. von Neumann. On regular rings. *Proc. Nat. Acad. Sci (USA)*, 22:707–713, 1936.

[34] J. von Neumann. Examples of continuous geometries. *Proc. Nat. Acad. Sci (USA)*, 22:101–108, 1936.

[35] R. Ware. Endomorphism rings of projective modules. *Trans. Amer. Math. Soc.*, 155:233–259, 1971.

[36] R. Ware and J. Zelmanowitz. Simple endomorphism rings. *American Mathematical Monthly*, 77:987–989, 1970.

[37] J.M. Zelmanowitz. Regular modules. *Trans. Amer. Math. Soc.*, 163:341–355, 1972.

Index

Frontiers in Mathematics

This series is designed to be a repository for up-to-date research results which have been prepared for a wider audience. Graduates and post-graduates as well as scientists will benefit from the latest developments at the research frontiers in mathematics and at the "frontiers" between mathematics and other fields like computer science, physics, biology, economics, finance, etc.

Advisory Board

Leonid Bunimovich (Atlanta), Benoît Perthame (Paris), Laurent Saloff-Coste (Rhodes Hall), Igor Shparlinski (Sydney), Wolfgang Sprössig (Freiberg), Cédric Villani (Lyon)

■ **Kasch, F. / Mader, A.**

Rings, Modules, and the Total

2004. 148 pages. Softcover. ISBN 978-3-7643-7125-8

In a nutshell, the book deals with direct decompositions of modules and associated concepts. The central notion of "partially invertible homomorphisms", namely those that are factors of a non-zero idempotent, is introduced in a very accessible fashion. Units and regular elements are partially invertible. The "total" consists of all elements that are not partially invertible. The total contains the radical and the singular and cosingular submodules, but while the total is closed under right and left multiplication, it may not be closed under addition. Cases are discussed where the total is additively closed. The total is particularly suited to deal with the endomorphism ring of the direct sum of modules that all have local endomorphism rings and is applied in this case. Further applications are given for torsion-free Abelian groups.

From the reviews:

"The book is self-contained, well organized and nicely written, making it a very effective introduction to the subject at hand: the total." (MAA REVIEWS)

Contents: Preface.- General Background.- I. Fundamental Notions and Properties.- II. Good Conditions for the Total.- III. The Total of Modules with LE-decompositions.- IV. The Total in Torsion-Free Abelian Groups.- Bibliography.- Index.

Frontiers in Mathematics

This series is designed to be a repository for up-to-date research results which have been prepared for a wider audience. Graduates and post-graduates as well as scientists will benefit from the latest developments at the research frontiers in mathematics and at the "frontiers" between mathematics and other fields like computer science, physics, biology, economics, finance, etc.

Advisory Board

Leonid Bunimovich (Atlanta), Benoît Perthame (Paris), Laurent Saloff-Coste (Rhodes Hall), Igor Shparlinski (Sydney), Wolfgang Sprössig (Freiberg), Cédric Villani (Lyon)